Sustainable Civil Infrastructures

Editor-in-Chief

Hany Farouk Shehata, SSIGE, Soil-Interaction Group in Egypt SSIGE, Cairo, Egypt

Advisory Editors

Khalid M. ElZahaby, Housing and Building National Research Center, Giza, Egypt
Dar Hao Chen, Austin, TX, USA

Sustainable Infrastructure impacts our well-being and day-to-day lives. The infrastructures we are building today will shape our lives tomorrow. The complex and diverse nature of the impacts due to weather extremes on transportation and civil infrastructures can be seen in our roadways, bridges, and buildings. Extreme summer temperatures, droughts, flash floods, and rising numbers of freeze-thaw cycles pose challenges for civil infrastructure and can endanger public safety. We constantly hear how civil infrastructures need constant attention, preservation, and upgrading. Such improvements and developments would obviously benefit from our desired book series that provide sustainable engineering materials and designs. The economic impact is huge and much research has been conducted worldwide. The future holds many opportunities, not only for researchers in a given country, but also for the worldwide field engineers who apply and implement these technologies. We believe that no approach can succeed if it does not unite the efforts of various engineering disciplines from all over the world under one umbrella to offer a beacon of modern solutions to the global infrastructure. Experts from the various engineering disciplines around the globe will participate in this series, including: Geotechnical, Geological, Geoscience, Petroleum, Structural, Transportation, Bridge, Infrastructure, Energy, Architectural, Chemical and Materials, and other related Engineering disciplines.

More information about this series at http://www.springer.com/series/15140

Hesham Ameen · Michele Jamiolkowski ·
Mario Manassero · Hany Shehata
Editors

Recent Thoughts in Geoenvironmental Engineering

Proceedings of the 3rd GeoMEast
International Congress and Exhibition, Egypt
2019 on Sustainable Civil Infrastructures –
The Official International Congress
of the Soil-Structure Interaction Group
in Egypt (SSIGE)

 Springer

Editors
Hesham Ameen
Housing and Building
National Research Centre
Cairo, Egypt

Mario Manassero
International Society
for Soil Mechanics
and Geotechnical Engineering
Turin, Italy

Michele Jamiolkowski
Polytechnic University of Turin
Turin, Italy

Hany Shehata
Soil Structure Interaction
Group Egypt
Cairo, Egypt

ISSN 2366-3405   ISSN 2366-3413   (electronic)
Sustainable Civil Infrastructures
ISBN 978-3-030-34198-5   ISBN 978-3-030-34199-2   (eBook)
https://doi.org/10.1007/978-3-030-34199-2

This Springer imprint is published by the registered company Springer Nature Switzerland AG
The registered company address is: Gewerbestrasse 11, 6330 Cham, Switzerland

Contents

About the Editors

Hesham Ameen Geo-Institute of HBRC is considered one of the largest geotechnical institutes in the Middle East that has an ISO certificate. Professor Hisham had been selected to chair the institute from many years ago. This selection had been done based on his activities, accuracy, awareness, and hard work in the research. He is specialized in geotechnical testing, neural networks for the geotechnical engineering, and soil Improvement.

Michele Jamiolkowski
Academic Records:
1971–1979, Associate Professor,
1980–2007, Full Professor of Geotechnical Engineering, Technical University of Torino.
Since 2008, Emeritus Professor of C.E. Technical University of Torino.
Founder and Chairman of the Engineering Consultant Company, Studio Geotecnico Italiano.

Most relevant appointments:
1985–2011, Geotechnical Consultant for the Suspension Bridge over Messina Straits.
1994–1997, President of the International Society for Soil Mechanics and Geotechnical Engineering.
1990–2001, Chairman of the International. Committee for Safeguard of the Leaning Tower of Pisa.
1996–present, Member of the International Advisory

Group of the European Bank for Reconstruction and Development for the design and construction of the New Safe Confinement of the reactor in the Chernobyl Nuclear Power Plant.

1992–present, Chairman of the International Board Expert for Development of the Second World Largest Copper Mine Tailings Depository Zelazny Most in Poland.

2001–2011, Geotechnical Consultant for the Engineering Company Technital designer of the MOSE Project in Venice for Safeguarding Venice from high tides.

2005–present, Chairman of the Technical Committee for Safeguard of Rome Monuments During Construction of the New Subway Line C Underpassing Historical Centre; 2005–present, Foreign Associate, US National Academy of Engineering.

Awards and Honors (selected):

K. Terzaghi and R. B. Peck Awards, from the ASCE.

De Beer Awards, from the Belgian Geotechnical Society.

Honorary International Member of the Japanese Geotechnical Society.

Doctor Honoris Causa: University of Bucharest, University of Ghent, SGGW, Life University (Warsaw).

Recipient of the Italian Prize "Savior of the Art".

Honorable International Member of the Japanese Geotechnical Society; 1998–present.

Honorary Professor Academia Sinica of Guangzhou, China.

Commendatore of the Italian Republic bestowed by the President of Italy.

Foreign Member of the Polish Academy of Science.

Member of the Lagrangian Academy of Science, Torino.

Mario Manassero obtained his Civil Engineering degree in 1980 at Politecnico di Torino and received his Ph.D. at the same university in 1987. He has been visiting professor at the University of Ancona (Italy) from 1988 to 1993, Ghent University (Belgium) in 1996, and at Colorado State University (USA) in 1995. Since 1998, he has been Professor of Geotechnical Engineering at Politecnico di Torino.

He has been Chairman of the Technical Committee (TC) no. 215 "Environmental Geotechnics" of the International Society for Soil Mechanics and Geotechnical Engineering (ISSMGE) for the period 2001 to 2014 and a member of the expert consulting board of the Italian Ministry of the Environment for the Environmental Impact Assessment of major national projects from 2008 to 2012.

His main research activities are devoted to the characterization of soil deposits by in-situ tests, soil improvement and reinforcement methods, containment systems for landfills and polluted subsoils, vacuum extraction of subsoil pollutants, and the mechanical behavior of municipal and industrial solid wastes. He has also addressed more fundamental topics like the chemo-physical interaction between pore fluids and the solid skeleton of active clays, the multiphase coupled flows, and the associated subsoil pollutant transport phenomena.

He has been invited lecturer in a number of international conferences and academic celebrations. Among them, it is worth to mention the State-of-the-Art Lecture on Environmental Geotechnics, at the Millennium Conference "GEOENG2000" jointly organized by ISSMGE, ISRM e IAGEA, Melbourne, Australia (November 2000).

He has been appointed as the second R. Kerry Rowe Lecturer by ISSMGE TC 215 and the Lecture was delivered at the 19th International Conference on Soil Mechanics and Geotechnical Engineering (ICSMGE), Seoul (Corea), 2017.

He was involved in many committees for the preparation of guidelines and regulations, at the national and international level, concerning civil engineering and environmental aspects and he was a member of the Italian Geotechnical Society Committee, AGI-UNI-SC7, for the National Application Norm of the Eurocode n. 7 "Geotechnical Design" (CEN).

As far as his professional activity is concerned Mario Manassero was involved in many landmark engineering projects such as the protection of the Venice Lagoon, the reclamation and rehabilitation of the Rome International Airport area, the stability assessment of the red mud tailing basin at Portoscuso (Italy), the pollutant containment diaphragm wall at Cengio (Italy), and the design of the Messina Strait bridge foundations and anchor blocs. He has also been geotechnical consultant of the Victoria State Environmental Protection Agency (Australia), contributing to the environmental planning for landfill locations as well as to the landfill design guidelines.

He has authored, co-authored, and/or edited five books and more than 150 technical and scientific papers in journals and conference proceedings.

Hany Shehata is the founder and CEO of the Soil-Structure Interaction Group in Egypt "SSIGE." He is a partner and vice-president of EHE-Consulting Group in the Middle East, and managing editor of the "Innovative Infrastructure Solutions" journal, published by Springer. He worked in the field of civil engineering early, while studying, with Bechtel Egypt Contracting & PM Company, LLC. His professional experience includes working in culverts, small tunnels, pipe installation, earth reinforcement, soil stabilization, and small bridges. He also has been involved in teaching, research, and consulting. His areas of specialization include static and dynamic soil-structure interactions involving buildings, roads, water structures, retaining walls, earth reinforcement, and bridges, as well as, different disciplines of project management and contract administration. He is the author of an Arabic practical book titled "Practical Solutions for Different Geotechnical Works: The Practical Engineers' Guidelines." He is currently working on a new book titled "Soil-Foundation-Superstructure Interaction: Structural Integration." He is the contributor of more than 50 publications in national and international conferences and journals. He served as a co-chair of the GeoChina 2016 International Conference in Shandong, China. He serves also as a co-chair and secretary general of the GeoMEast 2017 International Conference in Sharm El-Sheikh, Egypt. He received the Outstanding Reviewer of the ASCE for 2016 as selected by the Editorial Board of International Journal of Geomechanics.

Comparison of Geotechnical Characterization and Resistance Parameters Obtained from CPT Test and Conventional Laboratories in Fluvial – Lacustrine Soils

Lucero Amparo Estevez Rey[1]([⊠])
and Ibrahim Muhammad Elbatran[2]([⊠])

[1] Applied and Environmental Geoscience, Tubingen University,
Bogota, Colombia
luceroestevez@usantotomas.edu.co
[2] Earthquake Engineering and Structural Dynamics,
National Technical University of Athens, Cairo, Egypt
Ibrahim_el_batran@hotmail.com

Abstract. This paper shows the analysis of geotechnical information from recent updates of the interpretation of correlated variables from static piezocone penetration test (CPTU), and the validation of geotechnical characterization and resistance parameters by means of conventional laboratory test and MASW alignment techniques, checked within a framework of geological conditions studied from the execution of the project in a stratigraphy of silty and sandy clays of a lacustrine fluvial formation.

The obtained information will be analyzed though CPT test correlations, such as cone resistance (q_t), friction ratio (f_r) and pore pressuring during cone penetration (U_m), in order to generate a stratigraphic geotechnical characterization (identifying the type of soil from Robertson identification charts, 2009) and resistance parameters (provided as peak friction angle, soil sensibility, over consolidation ratios and undrained shear strength) from the identification of soil type in situ. The results will be verified with laboratory test with the aim of generate complete view of the studied soil, and in this way to obtain the characterization and resistance parameters from alternative methods for finite element modeling with less uncertainly of behavior from imposition of constructive loads. The study includes the analysis pf three CPT test performed ad 20 m depth or reject, defined from the stratigraphy previously found in the place.

1 Introduction

Currently, there are many methods for the evaluation of engineering parameters, terms of geotechnical parameters, geology, flow characteristics among others, using in situ test. The piezocone test is a fairly complete, repeatable and widely used method for detection of resistance parameters; is a fast and minimally invasive method to determine the mechanical and transport properties of soil types (Elsworth and Lee 2007) that measures tip resistance, friction resistance and pore pressure at every centimeter of

© Springer Nature Switzerland AG 2020
H. Ameen et al. (Eds.): GeoMEast 2019, SUCI, pp. 1–38, 2020.
https://doi.org/10.1007/978-3-030-34199-2_1

advancement of the sensor in the soil. From many correlations, it is possible to obtain these readings, identification parameters and resistance of the materials, with the need in the analysis and design of construction projects (Mayne 2009).

It is possible to carry out dissipation test with the sensor, whose results allow to knowing the soil flow and consolidation conditions. The rate of dissipation depends of the coefficient of consolidation, with in turn depends on the compressibility and permeability of the soil (Goujun et al. 2011). The parameters analyzed allow to knowing the OCR over consolidation ratio, with in fine soils is important to know to its mechanical behavior and its history of efforts.

The CPT test results depend of many factors, including soil type and characteristics, stress history and consequently presence of fissures and cracks. Those last, increase the *in situ* hydraulic conductivity even evaluated in a lacustrine deposit. Results of the CPT test can be complementary with usual exploration campaigns, with the aim of generating a coinciding or redundancy information atmosphere in a geotechnical characterization study.

This paper presents a case study of the interpretation of three CPT test carried out at depth of 20 m or soil reject, in an alluvial deposit composed of clays and silty sands, together with the verification of interpreted results of geotechnical characterization and resistance parameters in a comparison with conventional laboratory results, with the intention of verifying coincident information; with the aim of generating information obtained in a less invasive way to later use in finite element models and structural designs.

2   Geotechnical Conditions of the Site

The area where the essays were conducted is identified as an area of extensive alluvial deposits with the presence of clays, peat and silty – clay sands with thin levels of gravel. Analyzing the geotechnical plates of the place, it shows deposits of fluvial, fluvioglacial origin located along fluvial valleys, by depositional process.

Resident deposits of Quaternary age were identified, corresponding to Residual Alluvial Deposits (Qal) which area characterized by a study of an unconsolidated, sandy and silty material with scarce gravels; the sandstones are of various granulometries. The unit can be composed of predominantly fine materials such as clay and silt, which occurs when the deposit environment corresponds to the lower energy zones of the river studies.

3   Execution of in situ Test

Three static cone penetration tests were carried out, whose depth was around 20 meters or rejection, as shown in Table 1. The rest were executed with a VTR 9700 equipment. In order to obtain results in the same depth of the stratum, measurements were recorded from 1 m deep from the ground level, in each centimeter of advance.

Table 1. Depths of execution and water table

Essay	Depth	Dissipation depth	Phreatic level
CPT1	20.10 m	9.13 m	6.50 m
CPT2	18.90 m (reject)	3.20 m	8.00 m
CPT3	16.07 m (reject)	15.31 m	15.30 m

4 Results Obtained

A variety of normalized indices are available to the define the end bearing, frictional and pore pressure response recorded around the tip of and advancing piezocone (Elsworth 2005). In the Fig. 1 the interrelations of these parameters are shown.

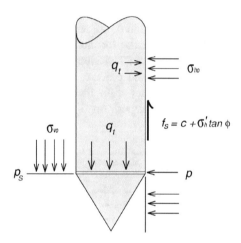

Fig. 1. Piezocone tip soil advance conditions. (Elsworth 2005)

4.1 Measured Variables

The variables read by the piezocone depend on the location and size of the pressure element, also influence the magnitude of the measured pore pressure specially in over consolidated stiff clays, where a larger pore pressure gradient develops around the tip (Abu – Farsakh et al. 1998). For the case study, the sensor is located behind the tip, as shown in Fig. 1, so it was possible to obtain the following data *in situ* (Douglas and Olsen 1891; Robertson et al. 1986):

- Cone resistance, Q_t
- Friction ratio, F_r
- Pore pressuring U_m

4.1.1   Correlations

The values should be interpreted from the use of the correlations established by area, unit weight, soil type, relative density, friction angle, dilatation angle, preconsolidation effort, undrained shear strength, lateral stress coefficient, modulus of elasticity, permeability, among other variables (Abu – Farsakh et al. 1998).

4.1.2   Correction by Area

For the case study, the following equations were used, using equipment calibration parameters:

$$q_t = q_c + (1 - a_n)u_2 \qquad (1)$$

$$f_t = f_s + b_n u_2 \qquad (2)$$

Where a_n and b_n are corrections by area calibrated in the laboratory. For the case study, these values correspond to 0.7 and 0.006 respectively.

The pore pressure u_2 depends on the location of the water level found in the borehole. Therefore, it is possible to establish input data: Normalized cone resistance q_t, Normalized sleeve friction f_s and pore pressure u_2 in units of force for each depth taken. Figures 2, 3 and 4 show the chars of the filed records, and Figs. 5, 6 and 7 the normalized records, for each of the test performed.

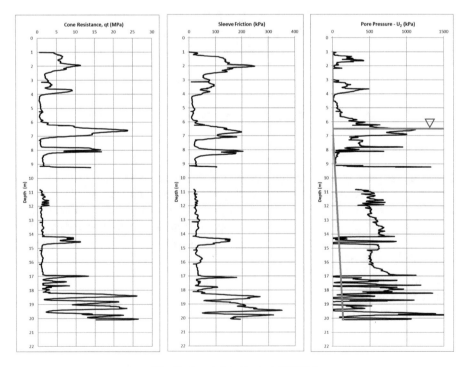

Fig. 2. Input data for test CPTU 1

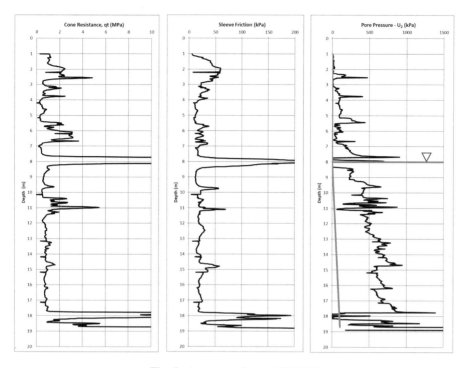

Fig. 3. Input data for test CPTU 2

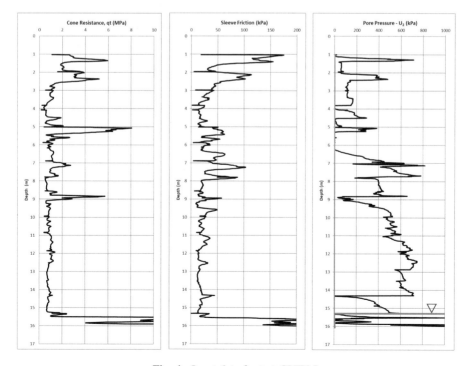

Fig. 4. Input data for test CPTU 3

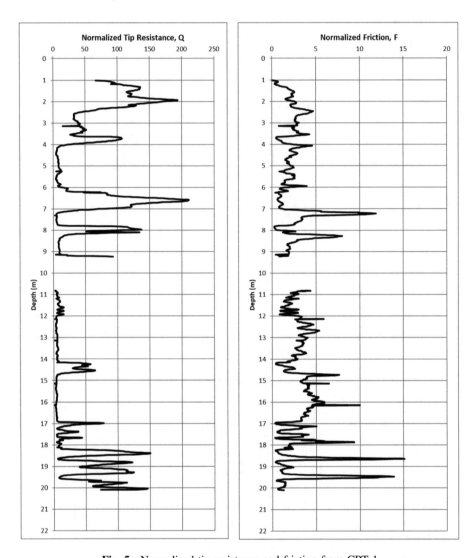

Fig. 5. Normalized tip resistance and friction from CPT 1

It is possible to corroborate with other methods of measured and/or obtaining variables, yet also plausible to find conflicting estimates where two approaches do not agree (Mayne 2004).

4.1.3 Soil Behavioral Type

The soil classification offered by the normalized CPT data (Normalized tip resistance, Q_t) is indirect, and should be corroborated by another test. It serves to give an estimate

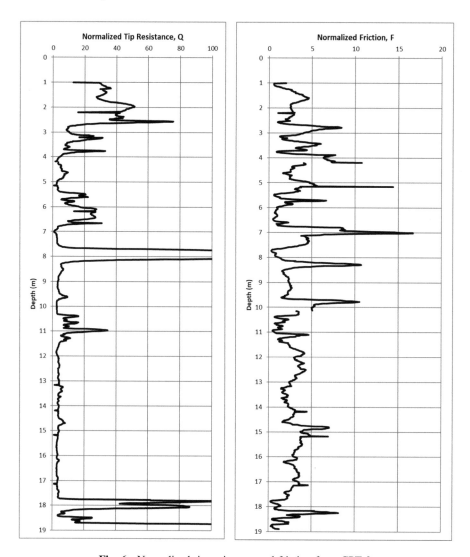

Fig. 6. Normalized tip resistance and friction from CPT 2

of the size of the material, but not the exact composition of its granulometry. It is possible to manage three approaches to correlate the type of soil:

– Rules of Thrumb (Mayne et al. 2002)
– Soil behavioral charts (Robertson 2009)
– Probabilistic methods (Tumay et al. 2008)

Each method is usable based on the type of geology of the area and the needs of the study. For the case study, "Soil behavioral charts (Robertson 2009) is adjustable since it can distinguish nine zones of different response, including fine soils, based on the

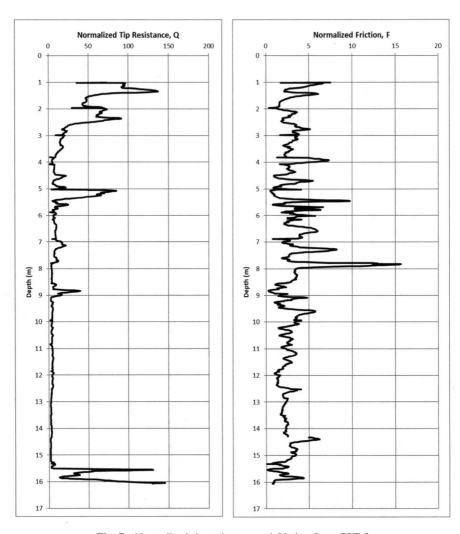

Fig. 7. Normalized tip resistance and friction from CPT 3

relationship between the correlation of the normalized tip friction and friction ratio, identifying I_c values.

$$I_c = \sqrt{(3.47 - LogQ_n)^2 + (1.22 + LogF_r)^2}$$ (3)

The graph Soil behavior type from $Q_{tn} - F_r$ chart with nine zonal classification is shown in Fig. 8.

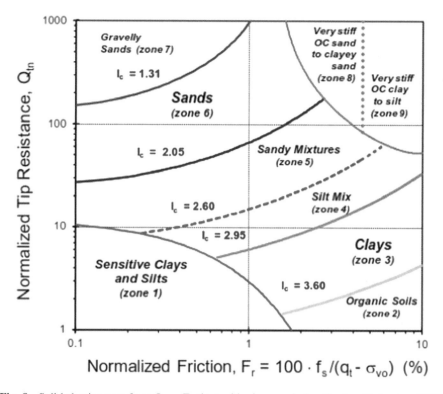

Fig. 8. Soil behavior type from $Q_{tn} - F_r$ chart with nine zonal classification (Robertson 2009)

The evaluation of the permeability in the soil can be carried out from the consolidation coefficient through the standard correlation with an independently evaluated coefficient of volume compressibility m_v, as it follows (Elsworth 2005):

$$K = \gamma_w m_v c_v \qquad (4)$$

Where γ_w is the water unit weight. In addition to the mechanical parameters mentioned in Eqs. (1), (2) and (4) the pore pressure can be defined as follows:

$$B_q = \frac{(u_2 - u_0)}{(q_t - \sigma_{vo})} \qquad (5)$$

This approach helps to identify soil types, especially those identified in zones 4, 5 and 6 of the Fig. 8, composed of sensitive or cemented clays. A chart of indirect classification based on soil behavior from $Q_t - B_q$ is shown in Fig. 9, as a result of empirical results proposed by Robertson (1990) and corroborated by Olsen (1994).

The results of the locations in the chars of each of the CPT test performed in the present study, using the aforementioned theories and methods, are shown in Figs. 10, 11 and 12. Table 2 shows the significance of the numbers that indicate a type of soil in the previous graphs (Fig. 13).

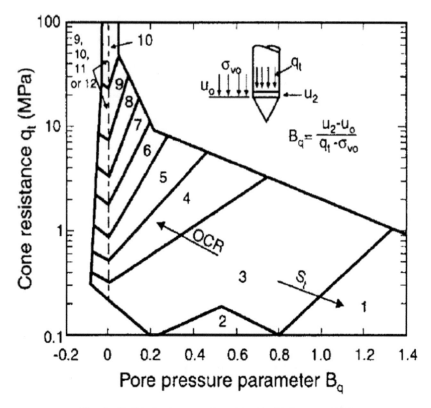

Fig. 9. Soil behavior type from Q_t - B_q (Robertson 1990)

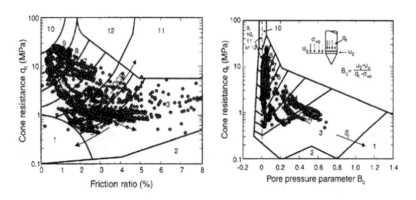

Fig. 10. CPT 1 results from Robertson et al. charts

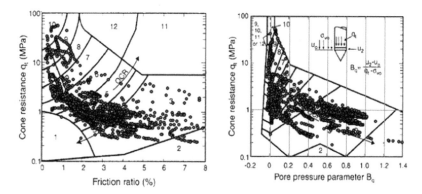

Fig. 11. CPT 2 results from Robertson et al. charts

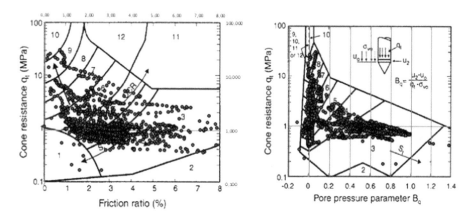

Fig. 12. CPT 3 results from Robertson et al. charts

Table 2. Type of soil related with charts (Robertson et al. 1986; Robertson and Campanella 1998)

Type	Soil behavior
1	Fine – grained sensitive soils
2	Organic material
3	Clay
4	Clay lime to clay
5	Clay silty to lime clay
6	Silty sand to lime sandy
7	Sandy line to lime sandy
8	Sand to sandy lime, over consolidated or cemented
9	Sands
10	Sandy gravels to sands
11	Very stiff fine grain soil
12	Sand to clayey sand

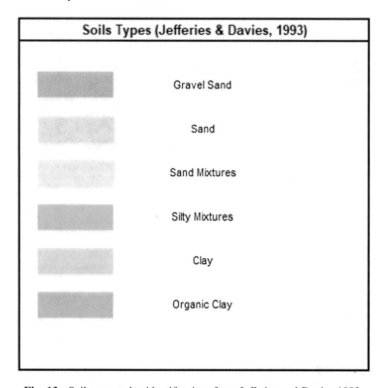

Fig. 13. Soil types color identification, from Jefferies and Davies 1993.

The results of classification index are also shown in the graph with color index, from graphing values of I_c (Eq. 3) with each depth value (Jefferies and Davies 1993). They are shown in Figs. 14, 15 and 16 for each essay performed; as also, the description of the stratigraphic profile obtained for each CPT test, corroborate with the classification information as a function of the friction angle and pore pressure parameter B_q, is shown in Tables 3, 4 and 5.

4.1.4   Soil Unit Weight and Shear Wave Velocity

It is possible to obtain the soil unit weight from calculations of s shear wave velocities, with a global relationship from Mayne (2007):

$$\gamma_t\left(\frac{kN^3}{m}\right) = 8.32 \, Log(V_s) - 1.61 \, Log(z) \tag{6}$$

Where V_s is calculated as follow:

$$V_s = \sqrt{\frac{\alpha V_s (q_t - \sigma v_0)}{P_a}} \tag{7}$$

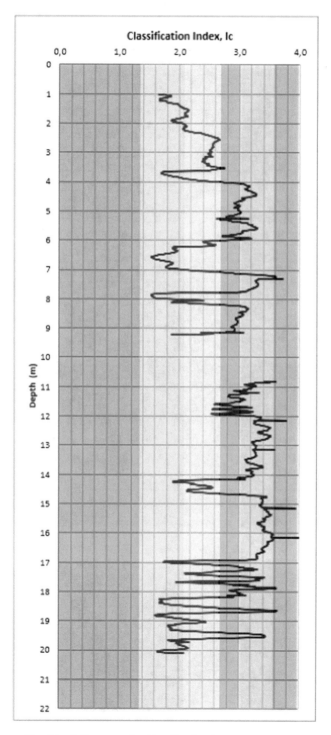

Fig. 14. Soil types color identification in depth for CPT 1.

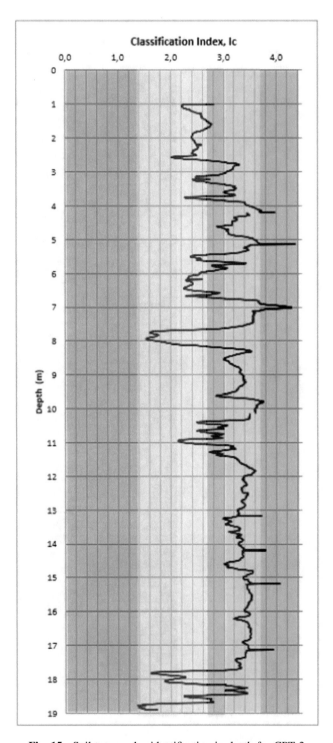

Fig. 15. Soil types color identification in depth for CPT 2.

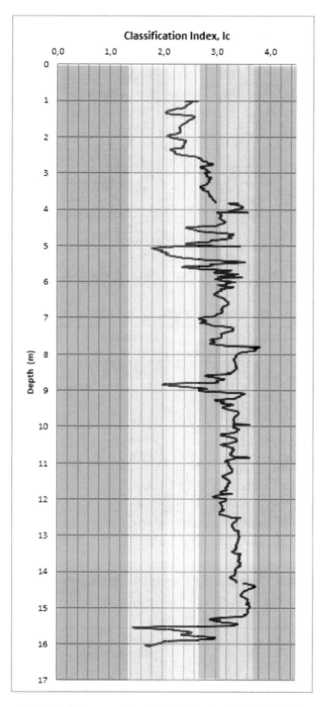

Fig. 16. Soil types color identification in depth for CPT 3.

Table 3. Description of the stratigraphic profile in the test CPT 1

Layer	Depth		Description
	From (m)	To (m)	
1	1.00	2.00	Sands
2	2.00	3.50	Sand mixtures
3	3.50	4.00	Sands
4	4.00	6.00	Clays
5	6.00	7.00	Sands
6	7.00	7.80	Clays
7	7.80	8.20	Sands
8	8.20	9.20	Clays
9	9.20	10.80	Sands
10	10.80	14.10	Clays
11	14.10	14.80	Sands
12	14.80	17.00	Clays
13	17.00	20.10	Sands & silts mixtures

Table 4. Description of the stratigraphic profile in the test CPT 2

Layer	Depth		Description
	From (m)	To (m)	
1	1.00	2.60	Sand mixtures
2	2.60	5.50	Clay – silt mixtures
3	5.50	6.70	Sand & silt mixtures
4	6.70	7.60	Clays
5	7.60	8.20	Sands
6	8.20	10.40	Claus
7	10.40	11.30	Silt mixtures
8	11.30	17.80	Clays
9	17.80	18.10	Sands
10	18.10	18.50	Clays
11	18.50	19.90	Sands & sands mixtures

Table 5. Description of the stratigraphic profile in the test CPT 3

Layer	Depth		Description
	From (m)	To (m)	
1	1.00	2.50	Sands
2	2.50	3.80	Silty mixtures
3	3.80	5.70	Sandy & silty mixtures
4	5.70	8.70	Clay & silty mixtures
5	8.70	9.10	Sands – sandy mixtures
6	9.10	15.50	Clay
7	15.50	15.70	Silty mixtures
8	15.70	16.80	Sandy & silty mixtures

Being αV_s a cone coefficient calculated from the recorded shear wave velocity, in units of velocity (Robertson 2009).

$$\alpha V_s = 10^{0.55 I_c + 1.68} \tag{8}$$

The graphical results of the mentioned correlations of unit weight and shear wave velocity are shown in Figs. 17, 18 and 19.

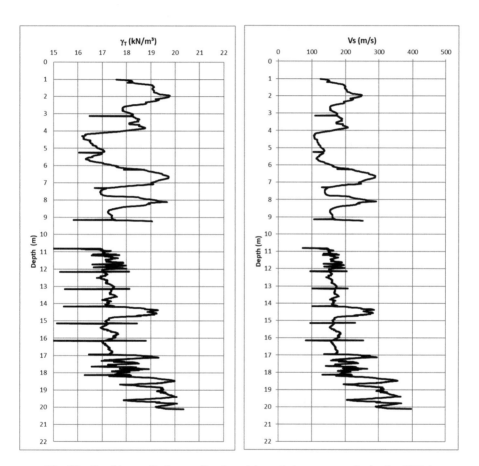

Fig. 17. Graphic results from soil unit weight and shear wave velocity for CPT 1

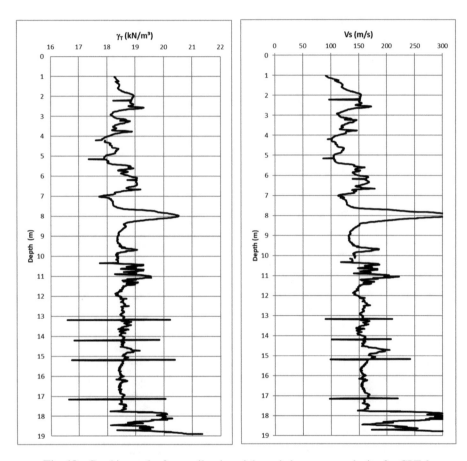

Fig. 18. Graphic results from soil unit weight and shear wave velocity for CPT 2

4.1.5   Friction Angle

The friction angle drained from the soils is an important parameter in the control as response of behavior of the loads and the initial stress state. The correlation used in this case study is based on an analysis of triaxial compression test derived by Kulhawy and Mayne (1990).

$$\varphi_p(^\circ) = 17.6 + 11.0 Log(q_{t1}) \tag{9}$$

$$q_{t1} = \frac{q_t - \sigma v_0}{\sigma_v'} \tag{10}$$

The plotted results of the friction angle correlation are shown in Figs. 20, 21 and 22.

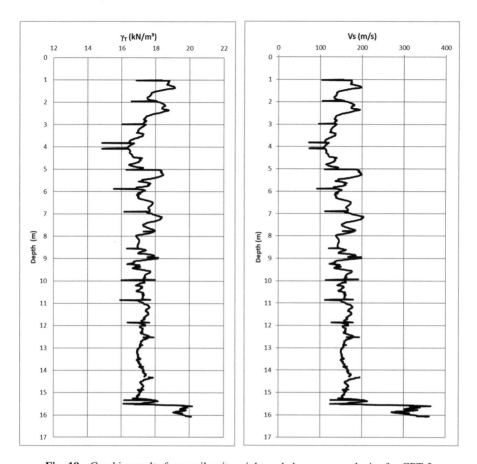

Fig. 19. Graphic results from soil unit weight and shear wave velocity for CPT 3

4.1.6 Overconsolidation Ratio – OCR

The over consolidation ratio determines the mechanical behavior of soft soils. It quantitatively describes the history of deposit stresses and determines the volumetric fluency condition in fine soils. In engineering, this value plays an important role in the calculation of shallow foundation settlements, and in some cases, it is used to estimate undrained strength parameters according to the correlations proposed by Skempton (2004) based on the preconsolidation value.

$$OCR = \frac{\sigma'_p}{\sigma'_{vo}} \tag{11}$$

$$OCR = 0.25 \left(\frac{q_t - \sigma_v}{\sigma'_v} \right) \tag{12}$$

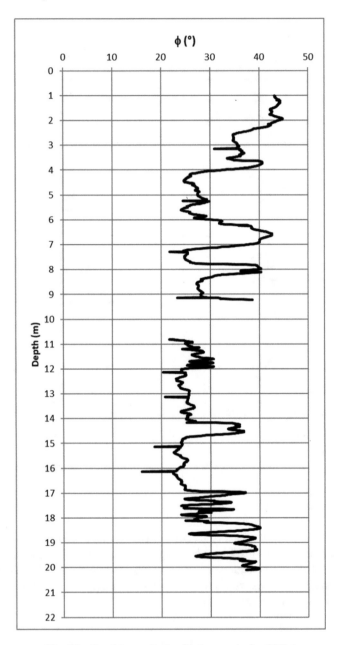

Fig. 20. Graphic results for friction angle for CPT 1

The graphical results of the over consolidation relationship with depth for each executed test are shown in Figs. 23, 24 and 25.

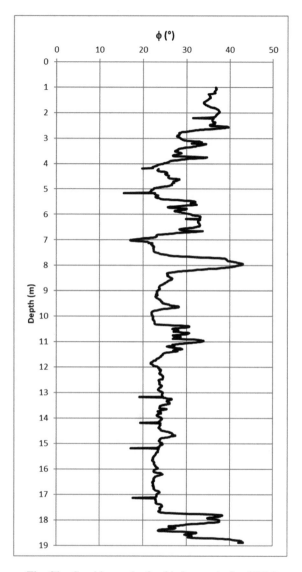

Fig. 21. Graphic results for friction angle for CPT 2

4.1.7 Passive Earth Pressure – K_o

The coefficient of passive earth pressure was calculated in relation with OCR and friction angle estimated by correlations. It is a medium parameter between the active and passive push of the material, which depends on the friction angle of the material.

$$K_0 = (1 - Sen\ \varphi)\sqrt{OCR} \tag{13}$$

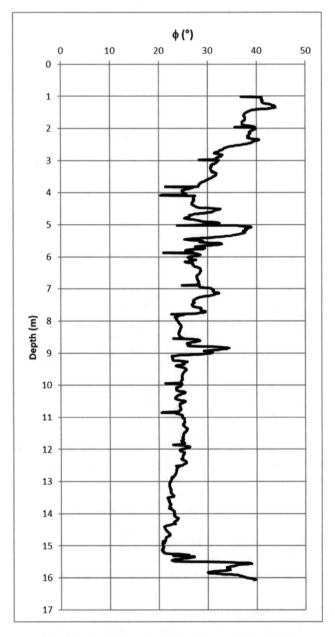

Fig. 22. Graphic results for friction angle for CPT 3

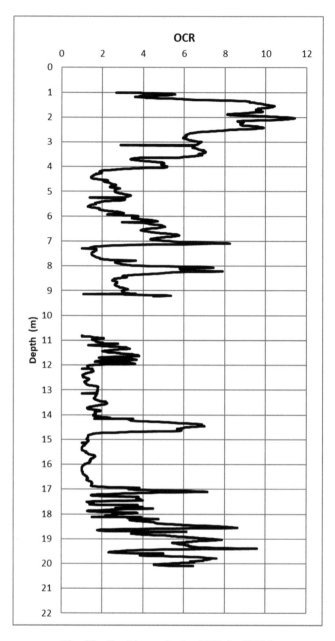

Fig. 23. Graphic results for OCR for CPT 1

4.1.8 Undrained Shear Strength

For normally consolidated fine-grained soul, the undrained shear strength ratio ranges from 0.2 to 0.3 (Jamiolkowski et al. 1985) with an average value of 0.22 in direct sample shear (Robertson 2009). New engineering tendencies are close to separate the

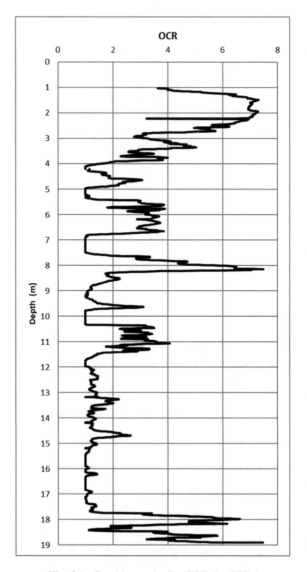

Fig. 24. Graphic results for OCR for CPT 2

influence of sensibility and OCR from the measured values of normalized cone resistance. Q_{tn} is not strong influenced by soil sensibility from $Q_{tn} - F_r$ chart (Robertson 2009). Based on those observations, and with a value of N_{kt} assumed of 14 for many insensitive fine-grained soils, the S_u correlation is:

$$S_u = \frac{Q_{tn}}{N_{kt}} \sigma'_{vo} \tag{14}$$

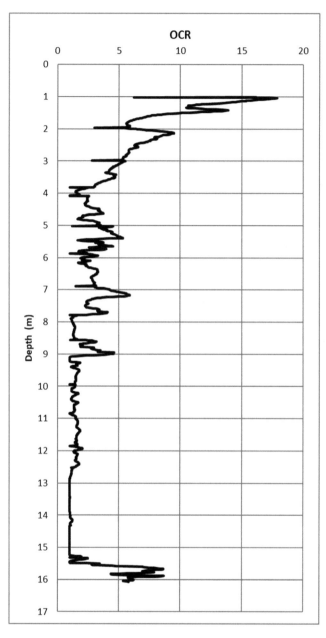

Fig. 25. Graphic results for OCR for CPT 3

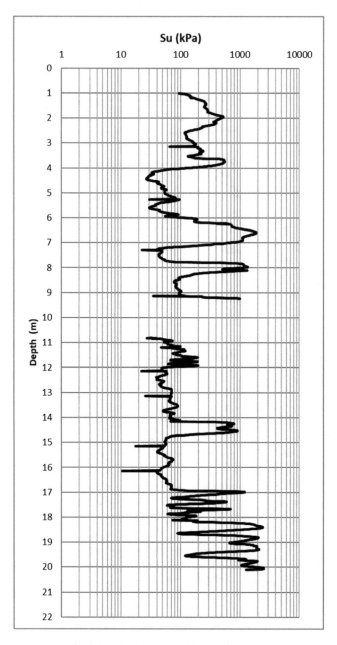

Fig. 26. Graphic results for Su for CPT 1

It is shown in the Figs. 26, 27 and 28 the correlation of undrained shear strength in a graphic with depth.

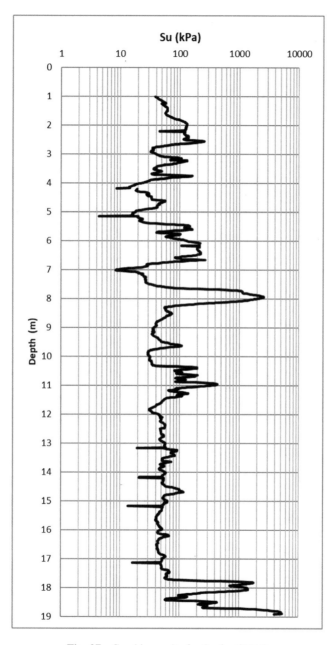

Fig. 27. Graphic results for Su for CPT 2

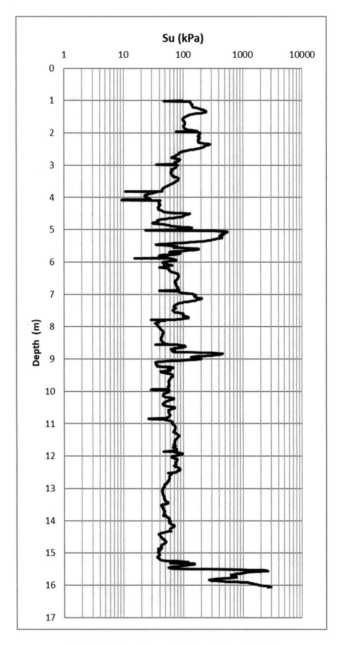

Fig. 28. Graphic results for Su for CPT 3

5   Analysis of Results

5.1   Stratigraphy Analysis

Reviewing the stratigraphy obtained from the soul behavior proposed by Robertson (2009), it is observed that the stratigraphic profile in the three CPT test is in accordance with Sand Mixtures – Silty Mixtures and Clays soil type, with some intercalations of organic material and sands. Analyzing the executed field test (SPT) and the laboratory results obtained with samples from perforations near the CPT locations, a stratigraphic profile composed of compact silty clays is obtained, with the presence of sands, of fine granulometry in fine upper percentage to 95%, with medium – high humidity; with plasticity indexes around 16–35; soil unit weights between 1.5 and 1.8 kN/m^3 and specific gravities around 2.4–2.6. Figure 29 shows the results of NSPT from boreholes near the location of CPT test, evidencing blows comprised between 5–20 mostly, and in Table 6 the laboratory results of the samples of related boreholes.

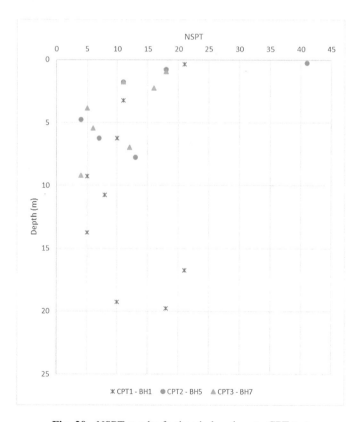

Fig. 29. NSPT results for boreholes close to CPT test

Table 6. Results of characterization laboratory test

Borehole	Sample	Depth (m)		Granulometry			Moisture %	Liquid limit	Plastic limit	Plasticity index	Shrinkage limit	Hydrometry and sieve	Soil unit weight	Specific gravity
Norm		From	To	% Gravel	% Sands	% Thin	NTC 1495 - 13	NTC 4603-99	NTC 4603-99	NTC 4603-99	NTC 1350-01	INV E 123-13	ISRM-07	INV E 128-13
BH1	3	1,50	2,20	0,00	0,90	99,10	25,37	54	19	35	19	99,1	1,883	2,673
	7	4,50	5,20	0,00	0,10	99,90	83,74	77	26	51	15	99,9	1,751	2,812
	11	7,50	8,20	0,00	0,10	99,90	32,68	50	18	32	18	99,9	1,866	2,861
	13	9,50	10,50	0,00	4,80	95,10	40,83	59	21	38	19	95,1	1,74	2,771
	16	12,00	12,60	0,00	3,00	97,00	37,79	39	15	24	9	97	1,857	2,683
	17	12,60	13,50	0,00	0,10	99,90	26,84	37	16	21	12	99,9	1,958	2,704
	20	15,00	15,70	0,00	0,20	99,80	29,80	46	20	26	13	99,8	1,82	2,746
	23	18,00	18,70	0,00	3,90	96,10	57,12	68	28	40	17	96,1	1,73	2,766
BH5	5	2,00	3,00	0,00	4,00	96,10	23,35	25	14	11	16	10,1	1,839	2,751
	6	3,00	3,70	0,00	10,90	89,10	16,39	46	19	27	6	96,9	1,969	2,729
	9	5,00	6,00	0,00	3,10	96,90	34,42	49	19	30	16	89,1	1,817	2,824
	14	9,00	9,80	0,00	89,90	10,10	191,25	282	128	154	18	96,1	1,538	2,677
BH7	6	3,00	3,60	0,00	0,30	99,70	30,44	43	20	23	13	99,7	1,883	2,685
	13	7,50	8,20	0,00	0,90	99,00	33,61	57	16	41	18	99	1,708	2,69
	14	8,20	8,90	0,00	1,90	98,20	37,21	46	22	24	14	83,2	1,873	2,664
	16	9,60	10,30	0,00	16,70	83,20	23,81	27	17	10	5	98,2	1,749	2,664

It is observed that the identification method based on correlations generates fairly close approximations that allow knowing the type of material in the area, with a stratigraphic profile composed of four materials was constructed:

Layer 1: Atrophic fillings: Conglomeratic sands of clast of up to 2 cm, on a sandy clay matrix, with traces of organic matter (0.00–1.30 m).

Layer 2: Silty clay materials: Composed mainly of predominantly clayey and silty materials with traces of grains of medium to fine sand, less than 3 mm in size (1.30–6.00 m).

Layer 3: Clay materials: Clay predominant materials, with traces of very fine sand, medium to hard consistency, medium moisture and plasticity (6.00–10.00; intervals 8.00–10.00).

Layer 4: Sandy Materials: These area materials that present sand content of medium to fine grain, with silty clay matrix, medium humidity and low plasticity (Intercalations 5.00–6.00 m, 10.00–20.00 m).

5.2   Parameters of Resistance and Rigidity

In terms of resistance, there are punctual increases of resistance in the time of piezocone penetration. For the CPT1 from the beginning of the drilling and up to 4.00 m, resistance is observed that oscillate between 2 and 7 Mpa with a peak at 2.00 and 3.75 m of 11 and 9 Mpa respectively. For CPT2, the resistance values vary between 1 and 2 Mpa. From 4.00 to 6.00 depth, the behavior for CPT1 and CPT2 is homogeneous, varying the tip resistance between 0.5 and 2 Mpa. From 6.00 to 9.00 m depth there is a significant increase in resistance per bit for CPT1 corresponding to sand lenses with values that fluctuate between 1 and 24 Mpa, however, for CPT2 the value varies between 1 to 3 Mpa. From 9.00 m to 17.00 m depth, the behavior of the tip resistance is similar with values that move in a range of 0.5 to 1.5 Mpa with some peaks of resistance corresponding to sand lenses at 9.00 and 14.50 m depth of 12 Mpa. After 17.00 m and until reaching rejection, the values fluctuate greatly due to the intercalation of sands and clays with values that start at 1 MPa up to 26 MPa.

About CPT3, a more homogeneous behavior is evidenced. From 1.00 to 2.50 m depth there is a variation in tip resistance that varies between 2 to 7 MPa. From this depth and up to 15.00 m depth the profile of resistance by tip presents a similar behavior with values raging between 0.54 to 1.50 MPa with some peaks at 5.50 and 9.00 m respectively. After 15 m and until rejection is reached, resistance values per point increase up to 28 MPa.

This information can be seen in more detail in Figs. 2, 3 and 4.

5.2.1   Friction Angle

The resistance parameters obtained by the aforementioned correlations, such as friction angle, are comparable by laboratory rest performed on the samples. Direct triaxial CU test were carried out, where the friction angle results are very close to those obtained by the CPT correlations in the three tests, fluctuating between 16 and 25°, which corresponds to soft materials and content of sands. Table 7 shows the result of laboratory test for friction angle, resulting from resistance test.

Table 7. Results of the friction angle taken from direct shear and triaxial essays

Borehole	Sample	Depth (m)		Direct shear (drained)	Undrained triaxial compression (TX - CIU)
		From	To		
Norm				INV E 131-13	INV E 153-13
BH1	11	7,50	8,20	11,3	
	17	9,00	9,70	22,1	22,1
BH5	10	6,00	6,60	32,2	
	12	7,50	8,10		23,3
	11	8,00	9,00		19,4
	14	9,00	9,80	28,1	
	20	12,00	12,70	27,9	
	24	15,70	16,50		19,6
BH7	6	3,00	3,60	31,5	
	11	6,00	6,70		26,52
	14	7,50	8,30	25,3	
	15	8,30	8,90		16,53

5.2.2   Shear Wave Velocity

The shear wave velocity calculated from correlations of the CPT for the soil unit weight calculation is comparable with V_s measurements taken from 8 kg impact hammer geophysical est. From the values of V_s obtained in the seismic lines it is possible to estimate the values of shear modulus and modulus of elasticity, assuming Poisson's ratio of 0.35 (clay materials). Because the elasticity values calculated from the V_s correspond to the maximum modules for low deformations, can not be used directly for settlement calculations, since the must be affected by a reduction factor to consider the levels of deformations that will be mobilized with the filling, and the admissible deformations depend of the factor of security that is chosen.

The results of the topographies interpreted by seismic refraction in the seismic refraction in the area where it is comparable with the CPT test results are shown in Figs. 30, 31 and 32. It is shown in the comparable with the CPT1 test, surface layer with oscillating shear wave values between 204 to 2016 m/s, corresponding to atrophic fillings and conglomeratic sands evidenced in the upper layers, as mentioned in the resulting stratigraphy. From this depth, the shear wave velocities decrease in magnitudes ranging from 192 to 312 m/s to the depth of interpretation of the test (approximately 25.0 m). Reviewing the values of the numerical correlation, values lower than 200 m/s are found in the upper depths, and since a depth of 6.00 m there is a gradual increase to 300–400 m/s. As far as can be concluded, it is a comparable result but not very accurate with the reality.

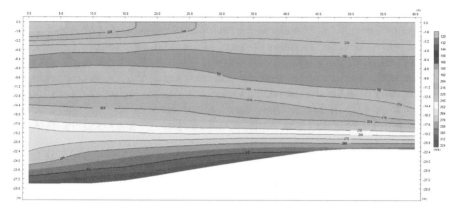

Fig. 30. Seismic tomography MSAW comparable with CPT1

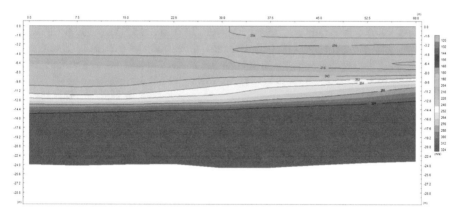

Fig. 31. Seismic tomography MSAW comparable with CPT2

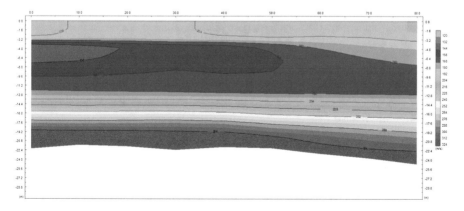

Fig. 32. Seismic tomography MSAW comparable with CPT3

Regarding the tomography comparable with CPT2, the shear wave velocity oscillates between 192 to 204 m/s in the first 3.00 m depth. A lens of probably sand material is evidenced that increases the value of the shear wave velocity to 216 m/s in the right margin of the profiled in the wave of 204 m/s. From this depth and up to 25.00 m depth, the shear waves velocities increase from 216 to 324 m/s. reviewing the values coming from the correlation, the values of the speeds oscillate between 100 and 150 m/s from 1.00 to 7.00 m, then they oscillate between 150 and 200 m/s until 17.00 m depth, where they later increase until more of 300 m/s gradually. This indicates that the trend of increase is shown in the correlation but not the nearby values, so it is not comparable with reality.

Finally, in terms of tomography comparable to CPT3, superficial layers with shear wave velocity values of 204 m/s are observed, corresponding to anthropic fillings and conglomeratic sands evidenced in upper layers. From this depth they decrease in magnitudes that oscillates between 156 to 180 m/s until 12.00 m depth corresponding to the clayey materials of the depth. From this point to the depth of the interpretation of the line, incremented values are presented between 180 and 324 m/s. Reviewing the correlated values, in depth comprised between 1.00 and 15.50 m, the shear waves velocities oscillate between 110 and 180 m/s, after that the gradually shoot up to 350 m/s until the rejection depth. In this case, the trend predicted by the CPT is not compatible with the data measured by geophysics.

The average calculated shear wave velocities are shown in Fig. 33. This result is important for future cases, a method of estimating modulus of elasticity both from correlated values and directly measured values.

### 5.2.3	Over Consolidation Profile

A calculation of settlements was carried out considering a law of logarithmic compression as a function of the parameters C_c and C_s measured by odeometric test, which requires the estimation of preconsolidation pressures. For the estimation of the over consolidation profile from SPT values, the correlation proposed by Mayne and Kemper (1988):

$$OCR = 0.193 \left(\frac{N}{\sigma'_v}\right)^{0.689} \qquad (15)$$

Performing the calculation of the over consolidation relationship from the results of the unidimensional consolidation test, it finds values close in trend and magnitude to the over consolidation ration calculated by correlations of the CPT. Table 8 shows the calculation information.

### 5.2.4	Undrained Shear Strength

Application of the undrained strength ratio can be useful since this relates directly to the OCR. The undrained shear strength for comparison, used in this article, was calculated from the friction angle obtained from laboratory test, the over consolidation ratio calculated previously and the initial vertical effort, data from characterization test.

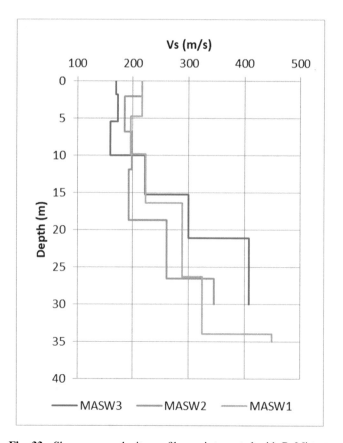

Fig. 33. Shear wave velocity profiles vs interpreted with ReMi test.

Table 8. Results of OCR calculation results from odometrical test

Borehole	Sample	Depth (m)		Consolidation test (Odeometer)	OCR
		To	From	σp (kPa)	
BH1	17	9,00	9,70	477,58	2,89
BH5	10	6,00	6,60	230,46	1,87
	14	9,00	9,80	120,00	0,83
	20	12,00	12,70	360,88	1,62
BH7	6	3,00	3,60	125,00	2,01
	14	7,50	8,30	130,00	0,90

Figure 34 shows the graphic representation of the values, which in comparison with the correlations made from each CPT test, the values shown by the calculation are very lower and in a different trend of increase to the graphs from CPT. It is possible that a more specialized correlation is needed to generate adjusted values.

6 Uses and Applications

The CPT essay is ideal to apply in soft soils, of lacustrine origin, as much as possible of young deposit formations (Quaternity), since their correlations generate less uncertainly. It is not possible to use in hard soils with the presence of rocks, cause the cell may suffer damages, and the pressure does not dissipate in the expected way. Because the essay as such does not generate a sample, it is advisable to accompany it with sample extractions methods such as the SPT validation information.

The growing demands of modern construction projects require everyday more specialized and complete studies that do not give space to uncertainly, and subsequent failure of structures. This is why the use of constitutive models that cover the largest amount of information collected. To analyze accurately the CPT results, it is necessary to have a numerical model that takes into consideration the large deformation of the soil around the tip, nonlinear soils behavior, the boundary condition with penetration, and the soil – piezocone interface friction.

With geotechnical information from conventional exploration methods, it is possible to define variables from Mohr Coulomb, Cam Clay, Hardening Soil, among other constitutive models, although it is usually expensive. This is why test with advanced exploration techniques such as the cone penetration test with pore pressure measurement represents a multiple and complete alternative to characterize the soil in a quickly and detailed way.

In highly compressible soft soils, selecting a constitutive models such as the Hardening Soil is suitable when using a CPT test, since by correlations it is possible to establish data such as Drained Triaxial rigidity E_{50}, discharge stiffness and reload E_{ur}, and rigidity tangent for E_{od} odeometric primary load of each of the soil strata, and in this way generate an approximation and future validation to the laboratory results. It is an alternative to generate studies coupled to the complex behavior of the soil through back analyses.

There are existing some relationships between cone resistance and Vs (or small strain shear Modulus Go), bur most were developed for either sands or clays and generally Young deposits (Quaternity). Pleistocene age souls had Vs values 25% higher than Holocene age soils (Robertson 2009). But often the age of the deposit is not always known in advance for most small low risk projects.

G_o, the small strain shear Modulus is determined from shear wave velocity, and then, is possible to obtain Young Modulus linked to the shear Modulus.

Several interpretation procedures have been used to analyze piezocone penetration and to evaluate some engineering soil parameters. The procedures are based usually on bearing capacity theories, or the cavity expansion theories (Abu – Farsakh et al. 1998). The cavity expansion theory cannot correctly model the strain paths followed by the soul element during cone penetration with the resulting stresses are not necessary in equilibrium. The piezocone penetration test results depend on many factors, some of those is the presence of fissures and cracks. That presence increases the in situ hydraulic conductivity and coefficient of consolidation.

Elsworth and Lee (2005, 2007) proposed and explicit equation for evaluating the in situ hydraulic conductivity of soils in horizontal direction from piezocone sounding records. However, the equation can be used reliably only for soils with a hydraulic conductivity typical of fine sand or soils with higher values of hydraulically conductivity. In clay soils, the main obstacle is that penetration process occurs in conditions very close to undrained, and does not involve significant pore water Flow within the soil near of the cone tip (Chai et al. 2011).

Since Elsworth and Lee studies (2004), a methodology is developed that potentially enables permeability profiles to be recovered from cone metrics recovered from CPT soundings. The technique requires that penetration be partially drained, resulting in the penetration induced tip local pore fluid pressures reflecting the competition between the penetration process that generates excess pressures. It must be controlled by the permeability of the surrounding soul, with the magnitude of the measured excess pore pressures therefore directly indexing permeability (Elsworth and Lee 2007).

7 Conclusions

The route for obtaining geotechnical information of CPT test in fine soils is presented in this paper. The method works quite well for characterizations, and to generate an idea of the trend of the resistance parameters and after comparing with laboratory test. It is highly recommended to compare the information obtained by conventional methods such as laboratory test, indirect explorations, among others.

Though the piezocone test it is possible to determine the classification of the stratigraphic profile of the soil according to its mechanical characteristics of behavior in the site, resistance and rigidity, as well as the pore pressures induced by the penetration. This stratigraphic profile, when compared to a profile studied from sample extraction and laboratory test, demonstrates a fairly accurate and even more complete approach, since it allows continuous measurements to be taken throughout the entire profile that reflect the behavior of the land without alterations.

In the study area, deposits of Quaternary age where identified, corresponding to Residual Alluvial Deposits (Qal) which are characterized by presenting unconsolidated, sandy and silty material with less grave bars; the sandstones found are of various granulometries (INGEOMINAS 2003). This unit can be composed of predominantly fine materials such as clays and silts, this happens when the deposit environment corresponds to the lower energy areas or the river currents.

The correlations mentioned in this paper, specially those for obtain resistance and rigidity parameters, generate more accurate results in deposits of soft and young lacustrine souls, that is, deposits of the Pleistocene and Holocene age (Quaternary).

The over consolidation relationship plays a fundamental role in the knowledge of the undrained shear ratio and settlements, so its calculation is essential in a geotechnical study using the available information.

The obtained results from geotechnical characterization by the correlations of the CPT test allow us to generate an approximate and very close view of reality, when their results are validated with laboratory test. This allows to determine with high reliability the basic parameters of a complex constitutive model, such as the Hardening Soil. Its limitations for the moment, are the determination of parameters such as cohesion and angle of dilatancy, which must be obtained by laboratory test.

From the CPT test and its correlations, it is possible to calculate different soil parameters such as the over consolidation ratio, undrained shear strength, friction angle, shear wave velocity, among others. These parameters allow a better understanding of the behavior of the soul, being parameters that are close in proximity to the value, even in tendency of increase and decrease in the depth. Similarly, for the estimation of stiffness parameters it is important to use the appropriate correlation, with a greater number of variables, in order to obtain a comparable and validate approximation with laboratory rest and indirect explorations.

Load settlement response for both shallow and Deep foundations can be accurately predicted using the measured shear wave velocity Vs. Although strong relationships among V_s and penetration resistance exist, some variability should be expected.

References

Robertson, P.K.: Interpretation of cone penetration test – a unfield approach, no. 46, pp. 1337–1355. NCR Research Press (2009). cgj.ncr.ca

Burns, S.E., et al.: Monotonic and dilatory pore – pressure decay during piezocone test in clay. Can. Geotech. J. **36**(6), 1063 (1998)

Elsworth, D., et al.: Limits in determining permeability from on – the – fly uCPT sounding. Geotechnique **57**(8), 679–685 (2007). https://doi.org/10.1680/geot.2007.57.8.679

Abu Farsakn, M.Y., et al.: Numerical analysis of the miniature piezocone penetration test (PCPT) in cohesive soils. Int. J. Numer. Anal. Methods Geomech. **22**, 791–818 (1998)

Elswoth, D., et al.: Permeability determination from on – the – fly piezocone sounding. J. Geotech. Geoenviron. Eng. ASCE (2005). https://doi.org/10.1061/(asce)1090-0241(2005) 131:5(643)

Elsworth, D., et al.: Dislocation analysis of penetration in saturated porus media. J. Eng. Mech. **117**(2), 391–408 (1991). ISSN0733-9399/91/002-0391. Paper No. 25504

Mayne, P.W.: Profiling OCR in stiff clays by CPT and SPT. Geotech. Test. J. **11**, 139–147 (1988)

Mayne, P.W.: Interpretation of geotechnical parameters from seismic piezocone test. In: Robertson, P.K., Cabal, K.I. (eds.) Proceedings of 3rd International Symposium on Cone Penetration Testing (CPT 2014, Las Vegas), ISSMGE Technical Committee TC 102, pp. 47–73 (2014)

Ballesteros Granados, R.V.: Obtaining parameters of the hardening soil model by means of CPTu test in soft soils of Bogotá. Polytechnic Magazine (2018). vol. 14, no. 26, pp. 89–97 (2014). (Spanish version)

Diaz Trillos, G.: Site characterization by means a CPTu test. Work degree to obtain a Civil Engineer Master. Pontificia Universidad Javeriana University, Bogotá, Colombia (2011)

Characterization of Shear Strength and Compressibility of Diesel Contaminated Sand

Sherif S. AbdelSalam[✉] and Ahmed M. M. Hasan

Civil and Infrastructure Engineering and Management, Faculty of Engineering and Applied Science, Nile University, Giza 12588, Egypt
{sabdelsalam, amahmoud}@nu.edu.eg

Abstract. Soil contamination with petroleum products and/or waste are a problem that can be detected nearby industrial areas and other amenities that include underground leaking tanks or pipelines. The negative effect of oil contamination on the soil properties is significant and can completely alter the strength as well as the serviceability limit states of the bearing stratum. In this study, Diesel was mixed with cohesionless soils using four different mixing percentages, starting with 5% up to 13.5% by weight, to cover a wide range of contamination ratios. The effects of contamination on the soil shear strength parameters and compressibility were assessed. Furthermore, the ultrasonic test was utilized to evaluate changes in the soil stiffness due to diesel contamination. Accordingly, a correlation was developed between wave velocity and California bearing ratio using frequencies ranging from 24 to 500 kHz.

Keywords: Diesel contamination · Sand · Direct shear · Shear strength · Compressibility · CBR · Ultrasonic on soil

1 Introduction

Contaminated soil is a hazard that affects not only the environment, but also the safety and integrity of foundations under new and existing structures. Risks of finding contaminated soils increase within industrial zones, nearby underground fuel tanks, heavy infrastructure facilities, or inland and offshore deep drilling locations (Shin and Das 2000). Most common soil contaminants that may be found at construction sites are formed of petroleum hydrocarbon substances such as diesel, gasoline, motor, and crude oils (Akpabio et al. 2017; and Safehian et al. 2018). Soil contamination also occur after environmental disasters such as the Gulf war (Al-Sanad et al. 1995; and Fine et al. 1997), which led to heavy contamination of sea water and soil with crude oil.

In general, the presence of such contaminant negatively affects shear properties of soil as indicated in previous studies such as those by Charkhabi and Tajik (2007); Nazir (2011); Kermani and Ebadi (2012); and Estabragh et al. (2014). However, not all outcomes from previous literature have indicated an unfavourable effect for the presence of oil products in soil, as some benefits may be achieved when a controlled, proper mixing of oil with soil take place. For instance, Al-Sanad et al. (1995) indicated that the

© Springer Nature Switzerland AG 2020
H. Ameen et al. (Eds.): GeoMEast 2019, SUCI, pp. 39–48, 2020.
https://doi.org/10.1007/978-3-030-34199-2_2

presence of specific types of lubricating agents enhance the sand properties and reduce its compressibility. According to Meegoda et al. (1998), benefits may include facilitation of compaction for sub-base layers during construction of road embankments, and enhancements in the soil shear strength and compressibility especially in compacted backfills used for ground improvement. This means that the bearing capacity of sand backfills may increase in the case of using some types of lubricants such as diesel Khamehchiyan et al. (2007). According to Abousnina et al. (2015), the California bearing ratio (CBR) increased by mixing soil with oil, and the cohesion increase by about 11 kPa for dry sand samples after adding 1% crude oil.

2 Laboratory Testing

Diesel is one of the most redundant contaminates that can be found in mixed with soil and ground water, which significantly alters the soil properties and behaviour. The laboratory-testing program aimed at characterizing the shear strength properties of the soil when mixed with Diesel, and these properties were mainly the angle of internal friction (ϕ) and the soil cohesion (c), as well as other compressibility indicators such as the CBR. The next sections of this study present the laboratory tests results conducted on the soil.

2.1 Classification and Diesel Percentages

The cohesionless soil adopted in the experimental program was clean course sand to represent the typical type of soil used for compacted fill. In order to determine the soil type, sieve analysis test was conducted where the soil was classified as poor graded sand (SP) following the unified soil classification system (USCS). In addition, a standard proctor test was conducted following the ASTM D698 in order to determine the maximum dry density (ρd_{max}) and corresponding optimum water content (OMC) of the sample. From Fig. 1, it was fond that ρdmax was about 1.92 gm/cm^3 at OMC of 6.5% and void ratio e = 0.26. This OMC and void ratio were used as basis for the water content and percentage of diesel that were determined and mixed with the soil.

Accordingly, three ratios of diesel were used as a contaminant and mixed with the SP soil samples. These ratios were 5%, 10% and 13.5% by soil weight. The contamination ratios were selected to provide comprehensive facts about the effect of diesel on soil shear strength and compressibility. The water content in each sample was constant and equal to 6.5%, knowing that the highest contamination ratio in the soil that is the 13.5% was selected in order not to exceed the natural void ratio of the uncontaminated soil sample (note that uncontaminated soil is namely in this study as the 0% diesel or the normal soil).

2.2 Direct Shear Test

The direct shear test (DST) was conducted on the SP soil samples using diesel contamination ratios 0%, 5%, 10%, and 13.5%. The DST was conducted following the ASTM D3080 (ASTM, 1997) to determine the soil shear failure envelop and

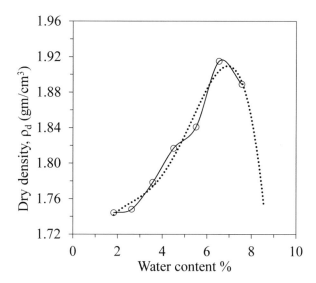

Fig. 1. Standard Proctor test for SP sample

accordingly determine ϕ and c, where both are the main soil parameters that influence on the ultimate bearing capacity of soil. The DST box dimensions were $100 \times 100 \times 20$ mm. For each test, three normal stresses (σ_n) were used and equal to 39.2, 78.5, and 157 kPa. The constant horizontal displacement rate was 0.5 mm/min, with a maximum displacement of 11.0 mm or until certain signs of failure appear on the load cell.

In Fig. 2, the load-displacement curves were plotted for each normal stress applied on the SP soil samples using the four percentages of diesel contamination. As can be seen from Fig. 2(a), the curves for samples with higher contamination tend to have lower shear stress values at failure, which also occurs at relatively larger horizontal displacements. Similar observations were noticed at higher levels of normal stress as presented in Figs. 2(b) and (c). The shear failure envelops of the SP soil samples at various percentages of contamination with diesel were plotted as shown in Fig. 3. This was done in order to determine the actual effect of diesel on the soil friction and cohesion. The figure shows a very slight (almost negligible) change in the friction angle ϕ, while some variations limited to ± 10 kPa were detected for the cohesion.

For a more in-depth breakdown, the exact percent of reduction in the friction angle and cohesion were plotted and presented in Fig. 4. It was confirmed from Fig. 4(a) that the friction values did not significantly change by increasing the diesel ratio, as ϕ was reduced by -6% at low contamination ratio of 5%, then actually started to increase again when the contamination ratio increased to 10% and 13.5%. Although the increase was limited to $+5\%$, but it means that using the diesel with certain percentages may actually enhance the friction properties of sandy soils. On the other hand, Fig. 4(b) shows the changes in the cohesion, which was a significant linear reduction that reached -50% with increasing the diesel ratio from 5% to 13.5%. The previous findings were investigated, and according to Khamehchiyan et al. (2007), this is

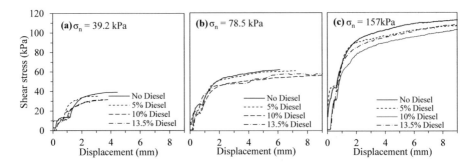

Fig. 2. Load-displacement curves from DST at normal stresses: (a) 39.2; (b) 78.5: and (c) 157 kPa

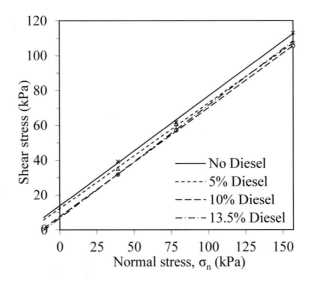

Fig. 3. Shear failure envelop for SP soil with diesel contamination

justified by the fact that the internal friction relies on the physical contact between sand particles, but cohesion relies on the surface contacts force between particles, which was reduced by the presence of a lubricating agents such as diesel and water in this case. It would also be important to indicate that the aforementioned changes in the ϕ and c values may significantly alter the soil bearing capacity; however may increase in case of pure sand with no cohesion.

2.3 California Bearing Ratio

The California bearing ratio (CBR) test was conducted on the contaminated SP samples to compare and validate the DST results, and to provide useful information for highway designers, as this test was originally developed by American Association of State

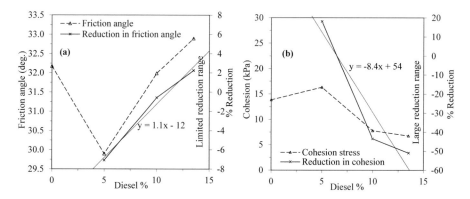

Fig. 4. Reduction in shear strength due to diesel contamination: (a) friction; and (b) cohesion

Highway and Transportation Officials ASTM D 1883 (ASTM 1997) to measure the quality of compacted soil layers under asphalt pavements. The test measures the penetration of a standard cylinder in soil compacted inside Proctor mold. The process of compaction is similar to the one used for standard Proctor test, whereas the SP soil with the same percentages of diesel contamination were used.

CBR load-penetration curves are shown in Fig. 5(a), where it can be seen that the load values tend to increase with the increase of diesel ratio up to a certain limit of about 10%. After that, the soil stiffness decrease again when the diesel ratio exceeds 10% up to 13.5%. This trend is similar to what can be achieved during a compaction test, which is clear as presented in Fig. 5(b). Accordingly, the figure shows that there is an enhancement in the CBR value from 5% to 25% that occur by increasing the diesel ratio to an optimum percentage of about 10% of soil weight, which also confirms with the DST results. Therefore there is a positive contribution of using limited amounts of diesel as a lubrication element during soil compaction, where stronger interlocking of between soil particles take place, leading to enhancements in the soil stiffness and compressibility. The was further investigated in the literature, and it was found that previous researches by Meegoda et al. (1998) and Abousnina et al. (2015) indicated that the of small amounts of lubricants to water during field compaction consumes less energy.

3   Ultrasonic Testing

The ultrasonic test was originally developed for concrete surfaces (Jones 1953; and Chamuel 1991) and then its use was relatively expanded to test other materials in the laboratory. Though, the use of this test on soils was very limited in the literature (Neilsen et al. 2019), with no evidence of adopting this test before for soils with diesel contamination. The device can be used to plot the transmitted and received waves, which gives an insight over how kinetic waves are transmitted through a certain mass – this is also essential in case of seismic loads. According to Neilsen et al. (2019), the

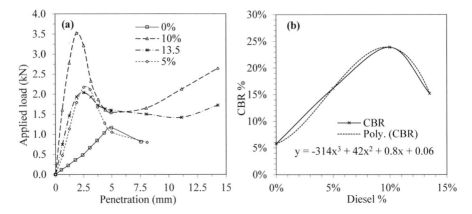

Fig. 5. CBR test for SP soil and diesel contamination: (a) load-penetration; and (b) CBR %

wave speed or velocity can give an indication about the soil modulus of elasticity, p- and s-waves, dynamic shear modulus, and soil damping coefficient.

The ultrasonic test was implemented in this study in an attempt to derive a correlation between the soil density and the speed of sound in the normal and diesel contaminated sand specimens, whereas the soil density or compactness can be linked with the previous CBR values. In the laboratory, the device used consisted of two probes for transmitting and receiving sound waves with frequencies ranging from 24 up to 500 kHz –Pundit device shown in Fig. 6(a). Figure 6(b) represents the steel box used during testing, where an open-ended steel box with internal dimensions equal to 70 × 70 × 70 mm was used. This steel box was used to contain the soil and to facilitate the transmission of sound waves through the tested soil sample.

Fig. 6. Ultrasonic test on diesel contaminated soils: (a) test device; and (b) used steel box

Figure 7 represents the wave amplitude versus time for the clean sand specimen (i.e., without diesel contamination) using different frequencies, 24 and 500 kHz. The phase shift between the two frequencies is clear on the figure, where the smaller frequency was relatively clearer inside the soil mass. Accordingly, the maximum amplitude of the 24 kHz wave was around 22.5% while that of the 500 kHz was limited to around 6%.

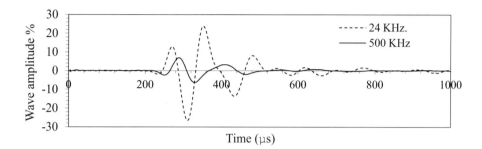

Fig. 7. Received signals for clean sand using different frequencies

Figure 8 shows the change in the wave velocity versus the percentage of soil contamination by diesel using different frequencies. From the Fig. 8(a), the effect of changing the frequency on the readings was evident, as the wave velocity using 24 kHz was slightly increasing, while vice versa in case of 500 kHz. The fringe between the transmitted frequencies could be due to the high damping properties of the contaminant. The amount of diesel in the soil added to the velocity variation. However, neglecting the first reading of the clean sand was considered to provide clearer outcomes. This is because removing the first reading avoids the bias of having two variables, which are water and diesel, as both have different acoustic properties. Therefore, it was better to acquire the relation between wave velocity and percentage of contamination by diesel as shown in Fig. 8(b). From the figure, an increase in the wave velocity was noticed by increasing diesel to a certain percentage, as the peak wave velocity ranged between 410 to 480 m/sec for 24 and 500 kHz, respectively, both at an optimum diesel percentage ranging from 8 to 10%. After that, the wave velocity started to decrease again by increasing the diesel percentage. From this figure, it can be observed that the change in the wave velocity follows a trend that is similar to that of the CBR test shown in Fig. 5(b), and that was evident for both frequencies.

Based on the previous, it was noticed that a correlation can developed between the wave velocity and the CBR (or soil stiffness/compressibility). This was developed and presented in Fig. 9. The points on the figure were developed using a wide range of diesel contamination using the two frequencies 24 and 500 kHz. From the figure, it can be seen that at a certain wave velocity there is a maximum and minimum CBR value that depends on degree of soil contamination. For instance, the wave velocity of 400 m/sec using a frequency of 24 kHz corresponds to a CBR value of 25%, while at a velocity of 475 m/sec using a frequency of 500 kHz corresponds to the same CBR value. In addition, the degree of diesel contamination has a significant effect on the CBR values as shown on the figure. Linear regression was considered to present a simplified cooperation between the wave velocity and CBR at various frequencies as

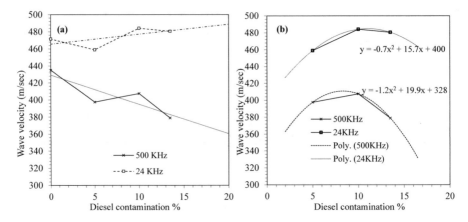

Fig. 8. Ultrasonic velocity versus diesel percent: (a) with clean sand; and (b) no clean sand

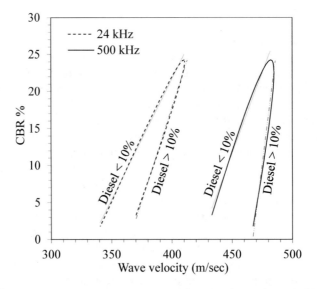

Fig. 9. Ultrasonic wave velocity versus CBR at frequencies 24 and 500 kHz

shown in the Eqs. 1 and 2 below, where V is the wave velocity (m/sec), and ω is the wave frequency (kHz).

For diesel contamination < 10%

$$CBR\% = \omega \left[2.3x10^{-4}V + 0.16\right] + 0.3V + 102 \qquad (1)$$

For diesel contamination > 10%

$$CBR\% = \omega \left[1.7x10^{-3}V + 0.9\right] + 0.5V + 160 \tag{2}$$

The aforementioned correlations can be used later to determine other soil parameters such as the modulus of elasticity after Putri et al. (2012), who provided Eqs. 3 and 4 below. In these equations, the C_u is defined as the ratio of uniform pressure imposed on soil to elastic settlement (δ) during CBR test, and p is the bearing pressure, E is modulus of elasticity, v is Poisson's ratio (0.3 for sand), and A is the area of the CBR load plunger.

$$C_u = p/\delta \tag{3}$$

$$C_u = 1.13\frac{E}{1-v^2}\frac{1}{\sqrt{A}} \tag{4}$$

4   Conclusions

A comprehensive laboratory characterization program was conducted on diesel-contaminated soil to determine the effect of contaminant on soil shear strength and compressibility. Tests such as direct shear, CBR, and ultrasonic were conducted on the sand samples using different percentages of contamination ranging from 5% to 13.5% by weight. A relationship was derived between the ultrasonic wave velocity and the CBR ratio using different frequencies equal to 24 and 500 kHz. The major findings of this study can be summarized as follows:

- Increasing the diesel contamination did not significantly affect the angel of internal friction of sandy soils. Change in the friction angle was limited to about 7% at diesel contamination percentage of 5%, and this change reduced to 2% with higher diesel percentages.
- Increasing the diesel contamination significantly affected the cohesion strength that is present in sandy soils, which dropped by about 60% at diesel percentage equal to 13.5%.
- CBR test was conducted to measure the change of sand compatibility in the presence of diesel contamination, whereas CBR increased with the increase of diesel contamination, and the peak value of the CBR occurred at 10% contamination. This highlighted the potential of using limited amount of diesel as an additive for ground improvement that can be effectively used during soil compaction.
- Ultrasonic tests conducted on contaminated sand samples showed that the wave velocity increase by increasing the percentage of diesel contamination. Results were also correlated with CBR outcomes and an equation was provided which can be used in the future to determine changes in the soil modulus of elasticity.

References

Abousnina, R.M., Manalo, A., Lokuge, W., Shiau, J.: Oil contaminated sand: an emerging and sustainable construction material. Procedia Eng. **118**, 1119–1126 (2015)

Akpabio, G.T., Udoinyang, I.E., Basil, T.S.: Effect of used motor oil contamination on geotechnical properties of clay soil on Uyo-Akwa Ibom. J. Nat. Sci. Res. **5**(2), 22–30 (2017)

Al-Sanad, H.A., Eid, W.K., Ismael, N.F.: Geotechnical properties of oil-contaminated kuwaiti sand. J. Geotech. Eng. **121**(5), 407–412 (1995)

Annual Book of ASTM Standards. ASTM International (1997)

Chamuel, J.R.: Ultrasonic studies of transient seismo-acoustic waves in bounded solids and Liquid/Solid interfaces (1991)

Chen, J., Wang, H., Yao, Y.: Experimental study of nonlinear ultrasonic behavior of soil materials during the compaction. Ultrasonics **69**, 19–24 (2016)

Estabragh, A.R., Beytolahpour, I., Moradi, M., Javadi, A.A.: Consolidation behavior of two fine-grained soils contaminated by glycerol and ethanol. Eng. Geol. **178**, 102–108 (2014)

Fine, P., Graber, E.R., Yaron, B.: Soil interactions with petroleum hydrocarbons: Abiotic processes. Soil Technol. **10**(2), 133–153 (1997)

Izdebska-Mucha, D., Trzciński, J., Żbik, M.S., Frost, R.L.: Influence of hydrocarbon contamination on clay soil microstructure. Clay Miner. **46**(1), 47–58 (2011)

Jones, R.: Testing of Concrete by Ultrasonic-pulse Technique. In: Highway Research Board Proceedings, vol. 32 (1953)

Kermani, M., Ebadi, T.: The Effect of Oil Contamination on the Geotechnical Properties of Fine-Grained Soils. Soil Sedim. Contam. Int. J. **21**(5), 655–671 (2012)

Khamehchiyan, M., Charkhabi, A.H., Tajik, M.: Effects of crude oil contamination on geotechnical properties of clayey and sandy soils. Eng. Geol. **89**(3), 220–229 (2007)

Meegoda, J.N., Chen, B., Gunasekera, S.D., Pederson, P.: Compaction characteristics of contaminated soils-reuse as a road base material. Recycled Mater. Geotech. Appl. 195–209 (1998)

Nazir, A.K.: Effect of motor oil contamination on geotechnical properties of over consolidated clay. Alexandria Eng. J. **50**(4), 331–335 (2011)

Neilsen, T.B., Robin, S.M., Sean, M., Margaret, G.M., Richard, S., Greg, A.V., Ingo, S., Andrew, G.H.: Preliminary analyses of seismo-acoustic wave propagation in outdoor field-scale analog volcanic explosions. J. Acoust. Soc. Am. **145**(3), 1869 (2019)

Putri, E.E., Rao, N.S.V.K., Mannan, M.A.: Evaluation of modulus of elasticity and modulus of subgrade reaction of soils using CBR test. J. Civil Eng. Res. **2**(1), 34–40 (2012)

Ratnaweera, P., Meegoda, J.N.: Shear strength and stress-strain behavior of contaminated soils. Geotech. Testing J. **29**(2), 133–140 (2006)

Safehian, H., Rajabi, A.M., Ghasemzadeh, H.: Effect of diesel-contamination on geotechnical properties of illite soil. Eng. Geol. **241**, 55–63 (2018)

Shin, E.C., Braja, M.D.: Bearing capacity of unsaturated oil-contaminated sand. Int. J. Offshore Polar Eng. **11**(3), 368–373 (2000)

Vasil'eva, A.A., Zobachev, N.M., Lobanova, G.L.: The use of ultrasonics for determining soil density. Soil Mech. Found. Eng. **6**(2), 95–98 (1969)

Utilization of Solid Waste Derivative Materials in Soft Soils Re-engineering

Kennedy Onyelowe[1,2](✉) (ID), A. Bunyamin Salahudeen[3],
Adrian Eberemu[4], Charles Ezugwu[5] (ID), Talal Amhadi[6] (ID),
George Alaneme[1] (ID), and Felix Sosa[7] (ID)

[1] Department of Civil Engineering, Michael Okpara University of Agriculture,
Umudike, P. M. B. 7267, Umuahia 440109, Abia State, Nigeria
konyelowe@mouau.edu.ng, konyelowe@gmail.com
[2] Research Group of Geotechnical Engineering, Construction Materials
and Sustainability, Hanoi University of Mining and Geology, Hanoi, Vietnam
[3] Department of Civil Engineering, Faculty of Engineering,
University of Jos, Jos, Nigeria
[4] Department of Civil Engineering, Ahmadu Bello University, Zaria, Nigeria
[5] Department of Civil Engineering, Faculty of Engineering,
Alex Ekwueme Federal University, Ndufu-Alike Ikwo, Ebonyi State, Nigeria
[6] Department of Construction and Civil Engineering, Ecole de Technologie
Superieure (ETS), University of Quebec, Montreal, Canada
[7] Department of Civil Engineering, National Autonomous University of Mexico,
Mexico City, Mexico

Abstract. Environmental degradation resulting from CO_2 emission and the constant collapse of foundation of facilities more especially pavements in Nigeria and across the world has posed serious threat to the overall economic growth of the developing nations. More so, Nigeria and the developing world lack an efficient solid waste disposal mechanism and policies hence indiscriminate disposal of waste on landfills poses yet another threat. The water resources in the developing countries is fast threatened by lack of engineered waste disposal facilities in different locations resulting to water pollution and its unhealthy consequences. This review work has brought to bear the interrelations between these problems. Geotechnical engineering in this paper promises to serve as a locus to bring these threatening environmental conditions into workable and beneficial stream. First, this paper tries to outline selected solid waste materials from which geomaterials utilized in the stabilization of soft soils, concrete production and asphalt modification are derived, by direct combustion or crushing. Secondly, the utilization of these derivatives, which serve as alternative cement in stabilization of soft soils, partial replacement for Portland cement in concrete production and modifier in asphalt production presents construction successes devoid of CO_2 emission because these materials are eco-friendly. Lastly, by adapting the use of these materials in soft soil, concrete and asphalt strength improvement, the solid wastes find a disposal path through the recycling process and eventual utilization as geomaterials, concrete additives and asphalt modification materials sources. Research results have shown that these materials derived from solid waste, because of their high

© Springer Nature Switzerland AG 2020
H. Ameen et al. (Eds.): GeoMEast 2019, SUCI, pp. 49–57, 2020.
https://doi.org/10.1007/978-3-030-34199-2_3

aluminosilicate content, improve the mechanical and strength properties of soils, concrete and asphalt.

Keywords: Solid waste reuse · Solid waste based geopolymer cements · Moisture bound materials · Geomaterials · Soft soils re-engineering · Asphalt modification · Concrete additives

1 Introduction

Disposal of waste in landfills poses a serious threat to the environment today. The water resources in the developing countries is fast threatened by lack of engineered waste disposal facilities in different locations resulting to water pollution and its unhealthy consequences. However, soft soils encountered in construction works pose a big threat to the overtime performance and operation of facilities whose foundation or underlain structure is made of these materials [1–6]. Pavements and pavement foundations are subjected to traffic loads, which could be axial or lateral pressure [1]. The structures suffer lots of unacceptable behaviour from construction cracks, volume changes to total collapse if the foundation is such that the material is weak and lacks the sufficient strength to bear the loads [2–4]. On the other hand, re-engineering of soft clay soils is a procedure or method adopted by the experts to improve the properties of soils to meet minimum standard requirements [6]. These minimum standard requirements are the conditions that justify the materials use as a foundation material or as a geomaterial [3]. Soil re-engineering involves mechanical, chemical and biochemical procedures. Through the application of these methods, soft soils achieve strength, density and durability [1–6]. In recent years, the problem of choosing between which of these methods gives a more sustainable approach has been there [4]. Earlier, Portland cement and other chemical additives and chlorides have been the only materials utilized as binders or pozzolans to improve on the mechanical properties of the soft soils or expansive soils [7]. Upon the utilization of these mentioned materials, there has been a build-up of the CO_2 emission into the atmosphere. This consequently contribute to global warming and eventually continue to put the environment at risk. For each tonne of Portland cement used in construction works, an equivalent amount of CO_2 is released into the atmosphere. Research has been ongoing to search for an alternative for Portland cement. It has been discovered that most ash materials derived from solid waste are pozzolanic in nature with very high content of aluminosilicates [7]. So also are most crushed materials into powder form. These ash and powder are all derived from direct combustion of solid waste materials [1]. For example, palm bunch ash, paper ash, periwinkle shell ash and snail shell ash, egg shell ash, bagasse ash, wood ash, rice husk ash, etc. are derived from palm bunch, waste paper, periwinkle shell, snail shell, egg shell, sugarcane fibre, wood, rice husk, etc. respectively [4]. Again, oyster shell powder, quarry dust, snail shell dust, sawdust, crushed ceramics, crushed glasses, crushed plastics, etc. are derived from oyster shell, quarrying, snail shell, wood sawing, waste ceramics, waste glasses, waste plastics, etc. respectively [1–6]. Research has shown that these materials contain high content of aluminosilicates hence highly pozzolanic [7]. This means that they can replace Portland cement as alternative cementing materials or supplementary cementing materials. They are all derivatives of solid wastes and sustainably serve the purpose of soft soil reengineering. A step further has been the synthesis of geopolymer cements with the derivatives of these solid waste materials as the base materials [8]. Pavement facilities fail every day and this is caused by the use of soils that are inadequate in strength and durability more especially in hydraulically bound conditions [9–13]. The aim of this work is to review the improvement that have been recorded in the utilization of geomaterials derived from solid waste in the re-engineering of soft clay soils with emphasis on; (i) review of selected geomaterials derived from solid waste recycling, (ii) review of the re-engineering process with these materials, and (iii) review of the improvements made or results achieved.

2 Methodology

2.1 Selected Solid Waste Materials

There have been lots of solid waste materials disposed into the environment more especially in the developing countries where solid waste management has been a huge problem. The selected solid waste materials from where the geomaterials are derived are palm bunch, quarry materials, palm kernel shell, snail shell, periwinkle shell, sugarcane fibre, waste ceramics, waste plastics, waste glasses, rice husk, wood, etc. These materials are solid wastes from industries, farms, homes, etc. and they are disposed indiscriminately into the environment. Geotechnical experts found need for these materials because of their aluminosilicate contents when burnt into amorphous ash or crushed into powder form [6]. The crushed materials most cases are biodegradable materials i.e. for those that are bio-based like snail shell powder, oyster shell powder, periwinkle shell powder. These are better when converted to ash by direct combustion. Oxide composition test conducted on these ash and powder materials showed that the alumina, silica ferrite composition was above 70% which is the minimum requirement for a materials to be considered pozzolanic [7]. The ash and powder materials are prepared and stored in bags. The methods of preparation are by direct combustion and crushing. And they are utilized in the stabilization of soft soils by measuring proportions in percentage by weight of solid and blended in matrix form with the treated soil. The number of proportions, which effects are to be examined determine the number of specimens to be prepared for the test procedure. In the case of biomass, ash or dust based geopolymer cement, the synthesis is enhanced under the reactive stimulus of Sodium Hydroxide (NaOH) and Sodium Silicate (Na_2SiO_3) which act as activators [1–6, 8–13]. These solid waste derivatives are all eco-friendly materials and with the utilization as a replacement for Portland cement in soft soil reengineering, CO_2 emission is reduced to zero. Within the mixed soil-additive blend, reactions take place with produce compounds responsible for strength gain and durability. This happens at the adsorbed complex or the double diffused layer (DDL). This is the interface of the dipole rearrangement where cation exchange reaction take place between the clay dipolar network and the additive dissociated ions. The formation of calcium aluminate hydrate (CAH) and calcium silicate hydrate (CSH) is of utmost importance because these are the compounds responsible for strength gain and durability.

2.2   Experimental Programs

Preliminary tests are conducted to determine the basic properties of the test soils. It has been observed that a good number of test soils collected for construction purposes don't meet the requirement to be considered acceptable for foundation works [9]. This is due to their formation, plasticity index, swelling potential, shrinkage limit, strength properties, etc. The results of these tests will determine whether the soil requires improvement, modification or re-engineering. Problem soils, that is, soils that are expansive, highly plastic, have high swelling potentials, have low shrinkage limits, with high cracking tendencies, weak, etc. require improvement to enhance the mechanical properties as to enable them meet the requirements to be used as foundation materials [14–19]. These treatment experimentations are carried out in accordance with British standard and Nigerian General Specifications [14–19]. The geomaterials utilized to improve the properties of soft soils are measured as proportions in percentage of the solid and mixed. Specimens are prepared according to the test being conducted, which of course determines the number of specimens to be prepared for test experiments. In very common cases, the additives are measure in an increment of 5% or 10% by weight of the solid. The effect of increased proportions of the additives on the properties is closely observed and recorded. General test methods are mostly used but in some cases some modified methods are adopted like in the case of durability test by loss of strength on immersion method, and stiffness test like resilient modulus and resistance value tests, which are modified methods of triaxial compression test.

3   Review of Results and Discussions

3.1   Compaction Characteristics

Research results have shown that the materials derived from solid wastes exhibit binding properties because of their aluminosilicates content and applied to soft soils as admixtures in a stabilization process improve the maximum dry density consistently with increased proportions of the additives [7, 8, 20–24]. This according results is due to the cation exchange reactions and the admixtures filling the voids within the soil matrix [59]. In addition, it is due also to the formation of floccs and agglomeration of the clay particles due to polarization, release and exchange of ions. The cause of this trend is because there was increasing desire for water, which makes up with the higher amount of admixtures. This is because more water was needed for the dissociation of the ions of admixtures with Ca^{2+} and OH^- ions to supply more Ca^{2+} for the cation exchange reaction [7, 20–24]. It was also observed that optimum moisture content reduced consistently with addition of these additives. This is due to the formation of floccs within the clay complex. The additives are highly pozzolanic materials and require water for hydration thereby improving the dry density development of the treated soils [20–24].

3.2 Consistency Limits

The utilization of geomaterials derived from solid waste has consistently caused an improvement of the consistency limits. Most of the soils we encounter in construction around here have a plasticity index of above 17%, which is highly plastic [7, 20–24]. The admixtures we have utilized in soils stabilization have improved the Plasticity Index (PI) from "highly plastic" soils to "medium plastic" consistency at the addition of the ash or powder materials. This trend showed that further addition of the admixtures will equally reduce the PI further [7, 20–24]. The hydration of the stabilized soil and its increased reactive surface has contributed to the improved behaviour of the soft soils and also due to molecular rearrangement in the formation of transitional compounds. This improvement is due to the hydration of the highly pozzolanic additives with the treated matrix, which reduced the PI consistently thereby producing a stiff mixture of stabilized soft soil. Also, the release of cations from the ash or powder materials and quarry dust during the cation exchange reaction has contributed to the enhanced behaviour of the treated soil. This behaviour agrees with Meegoda and Ratanweera [23], which showed that if water is used as pore fluid, the influence of the mechanical factors would remain same with a general decrease in LL on addition of any of the admixtures [7]. Consequently, the use of the treated soft soil as a subgrade and base material has been improved by the presence of the additives and this also achieved non-frost-susceptible materials with PI less than 15; a very important factor affecting the durability of pavements and other civil engineering works founded on soil [24]. The achieved subgrade improvement will reduce the required pavement thickness; wearing course + base course, hence a cost effective and more durable pavement construction.

3.3 Volume Change Characteristics

3.3.1 Swelling Potential

In hydraulically bound conditions like the pavement subgrade and the entire pavement, and other substructures constantly exposed to moisture attack, the foundation materials suffer changes in volume [7]. This is because the foundation materials are majorly clay or of high clay content like the problem soils we encounter in construction. When this underlain materials swell beyond the acceptable limits because of the rate at which moisture migrate to the clayey soil materials, they form heaves and cracks develop [20–24]. This cracks give way for more intake of moisture down the foundation and eventual collapse of the structure becomes evident. So, there is need to improve on the swelling behaviour of the foundation materials by admixture stabilization and of course the mechanical means by which densification is achieved. Research conducted has shown that these geomaterials utilized in the process of stabilizing soft clay soils improve the swelling potential of the treated soils to within acceptable limits [24]. This behaviour was due to that the higher content of sodium silicates activator for the case of solid waste based geopolymer cements and aluminosilicate in all other solid waste derivatives tend to increase the release of Ca^{2+}, Si^{4+} and Al^{3+} from the soft soil complex, which eventually speeded up geopolymerization and pozzolanic reaction rates. The Na_2SiO_3 activator acted as a nucleating site then increased with the amount of silicates released leading to the formation of more hydration points. And as the concentration of hydration materials increased, the number of contact points between hydration materials also increased consequently forming a solid microstructure within the treated soils matrixes reducing swelling potential.

3.3.2   Shrinkage Limits

On the other hand, when soft clay soils increase in volume as a result of water intake, it takes some time to revert to its original volume. In the case of clay, this reduction exceeds even the original volume to what we call shrinkage. There are acceptable limits for shrinkage problems in problem soils. When the soils shrink beyond the acceptable limit, then it leads to the formation of cracks. Crack formation is the major problem faced with use of Portland cement. But with use of biomass based geomaterials or geopolymer cements, crack resistant structure is achieved [7, 24].

3.4   Strength Developments Characteristics

3.4.1   California Bearing Ratio (CBR)

The CBR is the measure of a material to withstand shear failure by punching or penetration. It is an axial failure test conducted on soils to be used as foundation or subgrade materials. Various solid waste derivatives utilized as stabilizing agents have shown a consistent improvement in the CBR of the treated soft soils. These improved values > 20%, satisfy the material condition for use as improved sub-grade material on Nigeria's south eastern roads (Nigerian General Specification, 1997). The consistent increase in the CBR value with the addition of derivatives of solid waste was due to the presence of adequate amount of calcium required in the formation of Calcium Silicate Hydrate (CSH) and Calcium Aluminate Hydrate (CAH), which are the major compounds responsible for strength development [24]. The soil + additive blends passed to meet the minimum CBR value of 20–30% specified by Gidigasu and Dogbey [24], for materials suitable for use as base course materials when determined at MDD and OMC. Increase in CBR is an indication of the increase in MDD which is attributed to the compatibility of the grains of soils due to the increased reactive complex created by the additives and the high pozzolanic properties of the additives such that greater polycondensation and flocculation were achieved [7, 20–24].

3.4.2   Unconfined Compressive Strength

UCS is the measure of the ability of test materials to withstand compression. It is part of the stiffness measurement of construction materials like soils. During the treatment tests, the treated test specimens are cured for 7, 14 and 28 days. The test for durability was a modified form of the UCS test but in this case, the specimens were cured for 14 days and immersed for 14 days. The results of the additives effect on the UCS showed a consistent pattern of improvement in compressive strength. The presence of the admixture derived from solid waste materials in the soft soils increased the strength properties of the stabilized mixture. This is attributed to the physicochemical and highly pozzolanic properties of the admixtures from recycled solid wastes and to its ability to reduce adsorbed water thereby making soils with higher clay content to behave like granular soil. The addition of these additives derived from solid waste also improved the durability of the treated soft soils [24].

3.4.3 Resilient Modulus

This is one of the major stiffness determination factors that is a function of lateral pressure and deviatoric stress. The applied deviator stress and the recoverable strain techniques of the modified triaxial test on the treated specimens were used [7]. The tested soils behaved in almost the same pattern with similar reactions with increased additives of solid waste base. The deviatoric stress consistently increased with increase in the proportion of the admixtures for the test soils. It is important to note at this point that the additives are highly aluminosilicate compounds with a crystal texture prior to its utilization in the stabilization procedure [20–24]. These compounds are responsible for pozzolanic reaction, and strengthening by forming silicates of calcium hydrates and aluminates. Test soils had shown an improvement index of above 20. The highest improvement index recorded with test soils is in line with its natural soil high resilient modulus of $0.72E + 05$ which was improved upon. The hydration reaction between compounds of strengthening from the additives and the dissociated soil ions in contact with moisture had caused the improvement on both deviatoric stress and resilient modulus of the test treated soils. These results were recorded under cyclic loading on specimens subjected to testing sequences. The physical conditions that affect the resilient modulus (moisture and unit weight) were influenced by the introduction of the highly aluminosilicate admixtures derived from solid waste, hence improving the strength behaviour of the treated soils.

3.4.4 Resistance Value

Resistance value test conducted on solid waste base geomaterials treated soft soils showed a consistent improvement in the r-value of the soft soils increasing its resistance to lateral deformation [7]. This improvement was recorded with increased proportions of the geomaterials. This improvement was due to formation of CAH and CSH from the adsorbed complex reaction between clay and the additives [8, 24]. These compounds form the building block of strength gain and density development in treated soils thereby increasing its ability to withstand lateral deformation. This behaviour is due to molecular rearrangement between the dissociated ions from softs clay soils and the additives at optimum moisture within the DDL.

4 Summary Remarks

Geomaterials derived from solid waste, soft soil re-engineering, and results of the effect of solid waste derivatives on soft clay soils used as foundation materials have been reviewed. This article has outlined the methods of derivation of amorphous ash and powder from solid waste and the procedure through which these materials are utilized in the improvement of the mechanical and strength properties of soft soils. This is aimed at making the soft soils or problem soils or expansive soils usable as foundation materials in a stabilization process. This procedure not only improves on the properties of the soils to meet certain minimum standards for soils to be used as subgrade and subbase materials but also reduces to almost zero or zero the CO_2 emission into the atmosphere and also created an avenue for proper disposal of solid wastes through recycling. Hence the water resources in the developing countries that has been threatened by lack of engineered waste disposal facilities in different locations resulting to water pollution and its unhealthy consequences is yet restored to its healthy state.

Conflict of Interest. There are no conflict of interest recorded in this paper.

References

1. Onyelowe, K.C., Bui Van, D., Nguyen Van, M.: Swelling potential, shrinkage and durability of cemented and uncemented lateritic soils treated with CWC base geopolymer. Int. J. Geotech. Eng. (2018). https://doi.org/10.1080/19386362.2018.1462606
2. Onyelowe, K.C., Ekwe, N.P., Okafor, F.O., Onuoha, I.C., Maduabuchi, M.N., Eze, G.T.: Investigation of the stabilization potentials of nanosized-waste tyre ash (NWTA) as admixture with lateritic soil in Nigeria. Umudike J. Eng. Technol. (UJET) 3(1), 26–35 (2017). http://www.ujetmouau.com/
3. Onyelowe, K., Bui Van, D., Igboayaka, C., Orji, F., Ugwuanyi, H.: Rheology of mechanical properties of soft soil and stabilization protocols in the developing countries-Nigeria. Mater. Sci. Energy Technol. 2(1), 8–14 (2018). https://doi.org/10.1016/j.mset.2018.10.001
4. Onyelowe, K.C.: Solid wastes management (SWM) in Nigeria and their utilization in the environmental geotechnics as an entrepreneurial service innovation (ESI) for sustainable development. Int. J. Waste Resour. 7(282), 2 (2017). ISSN 2252-5211
5. Onyelowe, K.C., Maduabuchi, M.N.: Palm bunch management and disposal as solid waste and the stabilization of olokoro lateritic soil for road construction purposes in Abia State, Nigeria. Int. J. Waste Resour. 7(2) (2017). https://doi.org/10.4172/2252-5211.1000279
6. Bui Van, D., Onyelowe, K.C., Van Dang, P., Hoang, D.P., Thi, N.N., Wu, W.: Strength development of lateritic soil stabilized by local nanostructured ashes. In: Proceedings of China-Europe Conference on Geotechnical Engineering, SSGG, pp. 782–786 (2018). https://doi.org/10.1007/978-3-319-97112-4_175
7. American Standard for Testing and Materials (ASTM) C618. Specification for Pozzolanas. ASTM International, Philadelphia, USA (1978)
8. Bui Van, D., Onyelowe, K.: Adsorbed complex and laboratory geotechnics of Quarry Dust (QD) stabilized lateritic soils. Environ. Technol. Innov. 10, 355–363 (2018). https://doi.org/10.1016/j.eti.2018.04.005
9. Onyelowe, K.C.: Kaolin soil and its stabilization potentials as nanostructured cementitious admixture for geotechnics purposes. Int. J. Pavement Res. Technol. (2018). https://doi.org/10.1016/j.ijprt.2018.03.001
10. Onyelowe, K.C., Bui Van, D.: Durability of nanostructured biomasses ash (NBA) stabilized expansive soils for pavement foundation. Int. J. Geotech. Eng. (2018). https://doi.org/10.1080/19386362.2017.1422909
11. Onyelowe, K.C., Bui Van, D.: Predicting subgrade stiffness of nanostructured palm bunch ash stabilized lateritic soil for transport geotechnics purposes. J. GeoEng. Taiwan Geotech. Soc. (2018, in press). http://140.118.105.174/jge/index.php
12. Onyelowe, K.C., Bui Van, D.: Structural analysis of consolidation settlement behaviour of soil treated with alternative cementing materials for foundation purposes. Environ. Technol. Innov. 11, 125–141 (2018). https://doi.org/10.1016/j.eti.2018.05.005
13. Onyelowe, K.C.: Nanosized palm bunch ash (NPBA) stabilisation of lateritic soil for construction purposes. Int. J. Geotech. Eng. TandF (2017). http://dx.doi.org/10.1080/19386362.2017.1322797

14. AASHTO. Guide for Design of Pavement Structures. American Association of State Highway and Transportation Officials (AASHTO), Washington DC (1993)
15. AASHTO T 190-09. Standard method of test for resistance R-value and expansion pressure of compacted soils. American Association of State Highway and Transportation Officials, Washington DC (2014)
16. AASHTO T 307. Standard method of test for determining the resilient modulus of soils and aggregate materials. American Association of State Highway and Transportation Officials, Washington DC (2014)
17. BS 1377 - 2, 3, Methods of Testing Soils for Civil Engineering Purposes, British Standard Institute, London (1990)
18. BS 1924, Methods of Tests for Stabilized Soil, British Standard Institute, London (1990)
19. Nigeria General Specification/Federal Ministry of Works and Housing. Testing for the selection of soil for roads and bridges, vol. II (1997)
20. Onyelowe, K.C.: Nanostructured waste paper ash treated lateritic soil and its California bearing ratio optimization. Glob. J. Technol. Optim. **8**(220) (2017). https://doi.org/10.4172/2229-8711.1000220
21. Ai, C., Li, Q.J., Qiu, Y.: Testing and assessing the performance of a new warm mix asphalt with SMC. J. Traffic Transp. Eng. **2**(6), 399–405 (2015). https://doi.org/10.1016/j.jtte.2015.10.002
22. Hasan, M.M., Islam, M.R., Tarefder, R.A.: Characterization of subgrade soil mixed with recycled asphalt pavement. J. Traffic Transp. Eng. **5**(3), 207–214 (2018). https://doi.org/10.1016/j.jtte.2017.03.007
23. Meegoda, N.J., Ratnaweera, P.: Compressibility of contaminated fine-grained soils. Geotech. Test. J. **17**(1), 101–112 (1994)
24. Gidigasu, M.D., Dogbey, J.L.K.: Geotechnical characterization of laterized decomposed rocks for pavement construction in dry sub-humid environment. In: 6th South East Asian Conference on Soil Engineering, Taipei, vol. 1, pp. 493–506 (1980)

Oxides of Carbon Entrapment for Environmental Friendly Geomaterials Ash Derivation

Kennedy Onyelowe[1,2]($\boxtimes$) (ID), A. Bunyamin Salahudeen[3],
Adrian Eberemu[4], Charles Ezugwu[5] (ID), Talal Amhadi[6] (ID),
and George Alaneme[1] (ID)

[1] Department of Civil Engineering, Michael Okpara University of Agriculture,
P. M. B. 7267, Umudike, Umuahia 440109, Abia State, Nigeria
konyelowe@mouau.edu.ng
[2] Research Group of Geotechnical Engineering, Construction Materials
and Sustainability, Hanoi University of Mining and Geology, Hanoi, Vietnam
[3] Department of Civil Engineering, Faculty of Engineering,
University of Jos, Jos, Nigeria
[4] Department of Civil Engineering, Ahmadu Bello University, Zaria, Nigeria
[5] Department of Civil Engineering, Faculty of Engineering, Alex Ekwueme
Federal University, Ndufu-Alike Ikwo, Abakaliki, Ebonyi State, Nigeria
[6] Department of Construction and Civil Engineering, Ecole de Technologie
Superieure (ETS), University of Quebec, Montreal, Canada

Abstract. Environmental friendly, environmental efficient, and sustainable civil engineering works have been studied with a focus on utilizing the derivatives of solid waste recycling and reuse to achieving infrastructural activities with low or zero carbon emission. The direct combustion model, which is a solid waste incinerator sodium hydroxide-oxides of carbon entrapment model (SWI-NaOH-OCEM) developed in the cause of this research has achieved a zero carbon release. It has been shown that CO and CO_2 emission during direct combustion can be entirely entrapped during the derivation of solid waste based supplementary cementing materials used as replacement for ordinary Portland cement in soft soils re-engineering works. This is achieved via the affinity between sodium hydroxide and oxides of carbon which gives rise to baking soda. Geomaterial ash has been synthesized for use in soft soil re-engineering with no hazardous emissions. The overall assessment of the present review work has left the environment free of the hazards of CO and CO_2 emission. It was shown that these supplementary cementing materials derived from solid wastes improve the engineering properties of treated soft clay and expansive soils. It has been shown that solid waste recycling and reuse is a hub to achieving environmental friendly, environmental efficient and sustainable infrastructural development on the global scale.

Keywords: Oxides of Carbon Entrapment · Reuse of solid waste
geomaterials · Environmental friendly re-engineering · Environmental efficient
re-engineering · Sustainable re-engineering · Soft soils re-engineering

© Springer Nature Switzerland AG 2020
H. Ameen et al. (Eds.): GeoMEast 2019, SUCI, pp. 58–67, 2020.
https://doi.org/10.1007/978-3-030-34199-2_4

1 Introduction

Solid waste handling and management around the world and especially in the third world countries have contributed to the alarming amounts of carbon oxides emission and on the threat of global warming (Onyelowe et al. 2019; Onyelowe and Bui Van 2018a and b; Onyelowe et al. 2018; Onyelowe 2017a, b and c; Onyelowe and Maduabuchi 2017); Onyelowe et al. (2017). Lots of human activities are part of this negative contribution to environmental hazard including; construction activities that utilize Portland cement, industrial activities that release oxides of carbon and other volatile gases, agricultural activities that release biomass and biopeels, mechanical activities for example engine-fuel combustion, which release CO/CO_2, etc. (Onyelowe 2017a). The direct combustion or crushing, after they have been carefully sorted, presented in Fig. 1, of agricultural, household, municipal and certain biomass materials is not left out (Onyelowe and Bui Van 2018a; Onyelowe 2017c; Onyelowe and Okafor 2015; Ezugwu 2015; Eberemu et al. 2014). All these activities release oxides of carbon into the atmosphere, which contribute over 80% of the worlds volatile and hazardous emissions. More important to deal with in this present work is the recycling and reuse of solid waste materials by sorting, burning to ashes and the utilization of the product of this process (ash) in various ways in construction works (Bui Van et al. 2018; Ezugwu et al. 2015; Ezugwu 2015). In recent times, there have been strong calls for experts in engineering to work and exercise their expertise and practice towards environmental friendly and efficient and more sustainable operations and designs to protect our planet from the dangers of global warming (Ezugwu 2015; Eberemu 2013). Ordinary Portland cement as it is known, contribute equivalent amount of CO_2 emission into the atmosphere as it has found its use in almost every civil engineering construction exercise. It is also known that cement has its low ebb/performance in terms of resistance to cracking, durability issues (long-term exposure to moisture), brittleness, fire and heat resistance and sulfate resistance (Onyelowe et al. 2018b; Onyelowe and Bui Van 2018a and b; Bui Van and Onyelowe 2018; Eberemu et al. 2014; Osinubi and Eberemu 2010, 2008). On the other hand, solid waste materials derivatives like ash (amorphous) have been discovered as possessing properties that not only replace cement in parts or whole, but improve on the resistance to sulfate attacks, heat, fire, suction, capillary action, sorption, cracking, etc. of infrastructures where they are adapted (Onyelowe et al. 2018; Onyelowe and Bui Van 2018a, b; Osinubi and Eberemu 2019). Further on, these materials because of their high pozzolanic properties or high aluminosilicates contents have found use in the synthesis of geopolymer cements utilized in various proportions in soft soils re-engineering (Abdel-Gawwadm and Abo-El-Enein 2016; Akbari et al. 2015; Onyelowe et al. 2018b; Onyelowe and Bui Van 2018; ASTM C618 1978; Hamidi et al. 2016; Hariz et al. 2017).

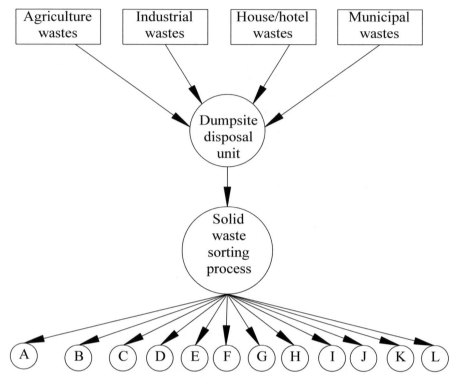

Fig. 1. Solid waste sorting process

2 Recycling and Reuse of Solid Wastes Methodological Review

2.1 Burnt Solid Wastes and Ashes

The greatest amount of ash materials utilized in soil stabilization, concrete production, asphalt modification and the synthesis of geopolymer cements (GPC) are derived through direct combustion in well-designed incinerators in a controlled burning process (see Fig. 2) (Onyelowe et al. 2018; Ikeagwuani and Obeta 2019c; Ezugwu 2015). In most cases, this combustion exercise is uncontrolled. This by implication poses the greatest danger of oxides of carbon emission to the environment (Vikas et al. 2018). In this case, a model of the incinerator has been designed to ensure that the oxides of carbon released through the firing smoke is entrapped. This mechanism is to ensure that the whole essence of an environmental friendly operation in civil engineering works is not defeated. The utilization of ash or its coupled forms in constructions and as geo-materials for the replacement of ordinary Portland cement is to reduce to zero emission, those construction practices that release oxides of carbon into the atmosphere to ensure an environmental friendly activity. But the production of ash by combustion has been against this aim. What has been done was to use the caustic soda-incinerator model to trap volatile gases released during the combustion process. The model is such that ensures that oxides of carbon (CO and CO_2) released during solid waste combustion is entrapped by caustic soda solution (prepared NaOH, 40% w/v). 100% of the CO and CO_2 released is trapped by the caustic soda solution because of the affinity caustic soda has with oxides of carbon. This entrapment produces sodium bicarbonate (baking soda, $NaHCO_3$), sodium carbonate (soda ash, Na_2CO_3) and hydrogen gas (H_2) presented in Eqs. 1 and 2;

$$NaOH + CO_2 \rightarrow NaHCO_3 \qquad (1)$$

$$NaOH + CO \rightarrow Na_2CO_3 + H_2 \uparrow \qquad (2)$$

The caustic soda (sodium hydroxide) entrapped oxides of carbon that would have been released to the environment with their dangers have been converted to household and industrial compounds. The use of baking soda ($NaHCO_3$) in households and industries cannot be overemphasized. So also is the use of soda ash (Na_2CO_3) in the synthesis of geopolymer cements; a geomaterial binder with great construction properties. Additionally, small amounts of the soda ash improves the flocculation properties of treated soft soils by improving the binding of fine particles. Hydrogen gas is also a product of the Solid Waste Incinerator NaOH Oxides of Carbon Entrapment Model (SWI-NaOH-OCEM). This gas and its isotopes have many uses in the field of science and atomic physics. It is important to note that the products of this model; SWI-NaOH-OCEM are water soluble compounds and element and find beneficial uses to man and the environment. However, ash, which is the primary product of the SWI-NaOH-OCEM has been used in several ways in construction works. Research results show that because of its amorphous nature and high aluminosilicates content, it acts as a binder with high pozzolanic property. It has in several cases consistently improved the engineering properties of expansive soils.

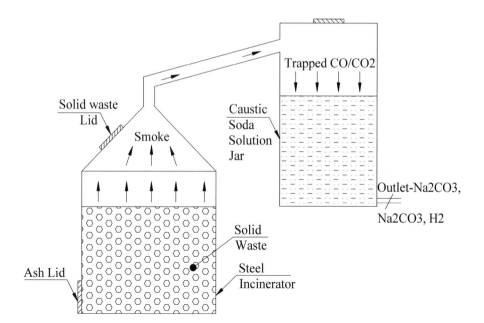

Fig. 2. SWI-NaOH-OCEM for solid waste combustion and CO/CO2 entrapment

3   Environmental Friendly, Environmental Efficient and Sustainable Re-engineering and Achievements of the New Tech

For far too long, the environment has suffered the hazardous effects of CO and CO_2 emission through human and engineering activities, which deplete the ozone layer and its attendant contribution to global warming. Experts of climate change in different for a, congresses and summits have given to the world what CO/CO_2 emission is doing to our planet (Onyelowe et al. 2019; Vikas et al. 2018). Players in this menace destroying our planet have been urged to seek for ways and alternative processes to renew our planet through environmental friendly activities and practices (Onyelowe et al. 2018; Onyelowe et al. 2019). The utilization of Portland cement in civil engineering works seems to contribute a large amount of the oxides of carbon to the atmosphere because research results have shown that ordinary Portland cement (OPC) use releases equivalent of amount of CO_2 into the environment, that is to say that one ton of OPC used in any construction activity release an equivalent one ton of CO_2 into the atmosphere. This is shocking and dangerous and if these activities continue, the future of this planet seems gloomy and uncertain. OPC is used in civil engineering work for the re-engineering of soils and concrete as structural and pavement foundation materials but this has failed both the ecofriendly dream environment by releasing CO_2 and eco-efficient product by failing durability tests because cemented structures are vulnerable to heat, sulfate, moisture, crack and shrinkage effects. On the other hand, ecofriendly materials that replace OPC as supplementary cementing materials are usually derived from direct combustion, another procedure that releases CO/CO_2 during the burning process. What this research has achieved with the ash derived from solid waste direct combustion, is trapping the CO/CO_2 released during solid waste controlled combustion using the SWI-NaOH-OCEM. From this model, we have the ash materials adapted to many different forms of geomaterials and construction materials, CO/CO_2 entrapment and environmental friendly products ($NaHCO_3$, Na_2CO_3 and H_2) for household and industrial uses.

3.1   Consistency Limits and Volume Changes (Swelling and Shrinkage)

Laboratory investigations carried out on soft clay soils treated with ash derived from solid waste and the coupled derivatives of ash; geopolymer cements and composites have shown tremendous and acceptable consistency behavior (Onyelowe et al. 2019). The properties of these amorphous materials utilized in the treatment protocol promotes the pozzolanic and hydration reactions between clay minerals and dissociated ions of the additives within the activation complex interface improving on the liquid limits, plastic limits and plasticity indexes of the treated soils. There are also indirect methods such as oedometer and free swell tests used to monitor the volume changes and consolidation settlement of these soft samples. The X-ray diffraction analysis (XRD), X-ray fluorescence (XRF), differential thermal analysis (DTA), dye adsorption, chemical oxides analysis, and scanning electron microscopy (SEM) belong to the mineralogical identification methods used to identify oxides and minerals contained in a soil specimen as to study the behavior when in contact with additives in a treatment procedure (Onyelowe et al. 2019). Although the mineralogical identification methods are capable of adequately recognizing the clay minerals in soft clay soils, they are somewhat restricted in use when characterizing the swelling behavior due to various technical setbacks. The identification and characterization encourages a better understanding of the anticipated behavior of the soils when treated with ash and its derivatives as discussed above. Results have also shown that these understanding has helped in the consistent reduction in the plasticity indexes of soft clay soils treated with varying proportions of powder, ash and ash or powder based geopolymer cements. Interestingly, these achievements have been under environmental friendly activities i.e. devoid of ordinary Portland cement utilization, these test soils have improved from very high consistency (PI > 17) to medium consistency (PI < 17) and to low consistency (PI < 7) meeting the consistency requirements of soil materials to be used as foundation materials (Onyelowe et al. 2019; Amadi and Eberemu 2012). Swelling and shrinkage properties are very important factors that influence the behavior of engineering soils as foundation materials especially in hydraulically bound environments. Soils as foundation materials suffer moisture attacks through capillary action, suction, rise and fall of water table and lateral percolation (Onyelowe et al. 2019; Onyelowe et al. 2018; Ikeagwuani and Nwonu 2019; Osinubi et al. 2017; Osinubi et al. 2012; Onyelowe and Ubachukwu 2015; Onyelowe and Okafor 2015). This is more prevalent in pavement structures. Volume changes cause undesirable characteristics that impair the performance and long-term service life of foundations and pavement infrastructures. Experimental results have shown that these undesirable properties of soft clay soils have been arrested by the addition of ash or powder or ash/powder based geopolymer cements in a stabilization procedure. First, the moisture resistance ability of the ash or powder materials derived from solid waste due to their composition and the hydration structure makes it possible for soils treated with them to behave within acceptable swelling limits (Onyelowe et al. 2018). There have been recorded reduction in swelling potentials through the addition of these additives. Conversely, these materials possess the ability to resist drying during the reverse phase of swelling i.e. reduction of entrapped moisture. Shrinkage in soft clay soils leads to cracking and lateral deformations but the additions of solid waste derivatives has changed this undesirable behavior due to the ionic composition of aluminosilicates admixtures.

3.2 Compaction, Strength Properties (CBR and UCS) and Durability

Compaction behavior of soft soils treated with additives derived from solid waste, ash is a density/moisture behavioral study. This translates to the densification achieved in soils during stabilization. This is a commonly used method of soft soils treatment achieving densification by mechanical methods and exerting compactive efforts and eventually reducing the voids in treated additive/soils blends (Amadi and Eberemu 2012; Eberemu and Osinubi 2010). This achieved to enable engineering soils withstand subsequent load without suffering immediate compression and this can be separated from those behavior characteristics initiated by long-term consolidation of soft clay soils. Ash materials are amorphous and contain ions dissociated when they react with clay minerals and water in a cation exchange reaction giving rise to ion combinations that support strengthening. The crushed materials (powder/dust) also act as fillers to improve on the porosity of the treated samples thereby improving the density. Results have shown significant reduction of the optimum moisture content (OMC) of treated soils more especially with the ash or powder based geopolymer cements because of the improved hydration reaction and moisture resistance of the polymers (Onyelowe et al. 2019; Onyelowe and Bui Van 2018a and b; Onyelowe et al. 2018; Onyelowe 2017a, b and c; Onyelowe and Maduabuchi 2017). The maximum dry density (MDD) obtained at OMC has also consistently improved in its properties under the influence of ash, powder and geopolymer cement additives in varying proportions. The intergranular pores are improved upon by the addition of ash or ash based geopolymer cement, which eventually encourages densification during compaction exercises. Shape of soil grains and amount of type of clay minerals present in treated soils are soil dependent factors that influence the behavior of soft clay soils during stabilization operations (Onyelowe et al. 2019; Onyelowe and Bui Van 2018a and b; Onyelowe et al. 2018; Onyelowe 2017a, b and c; Onyelowe and Maduabuchi 2017). However the addition of these additives has improved the interface behavior of these factors. California bearing ratio (CBR) is a strength characteristic feature of soils used to determine its suitability to withstand shear deformations by punching and penetration. It establishes the overall thickness of pavements and other horizontal infrastructures. Because this has a direct relationship with the density of compacted soils, MDD is directly related to CBR. Natural and treated compacted soils are subjected to axial load and the axial strains are monitored. There are two conditions under which this is experimented. These are the soaked and un-soaked bearing ratios determined under 2.5 mm and 5.00 mm penetrations. As discussed earlier on compaction behavior of the treated soft soils, the ash materials and its derivatives improve on the density properties giving compactness that determines the California bearing ratio under axial loads (Osinubi et al. 2012; Ikeagwuani and Nwonu 2019; Onyelowe et al. 2019). These additives improve on the soils ability to form a densified mass and compaction completes the process of densification and strengthening. Compressive strength of soft soils is one the strength characteristics of that has equally allowed the determination of long-term performance and behavior of treated soils more especially under hydraulic influence through loss of strength on immersion method. The failure point of a treated and compacted sample under compression is determined with respect to the sample surface area. The formation of calcium silicate hydrates and calcium aluminates hydrates with addition of ash, powder and ash/powder based geopolymer cements in soft soils promotes strengthening of treated soils. This is due to those hydrates responsible for strengthening and this improves the compressive strength of the treated samples. The innovation here is that these hydrates of strength are formed utilizing ecofriendly materials derived from solid waste hence with zero release of carbon oxides to the environment (Onyelowe et al. 2019). The durability potential is the compression index between completely open air cured samples and partly open air cured and partly full immersion cured specimens. As has been captured previously, these amorphous materials, powder and geopolymer cements have shown to resist moisture influence by reducing its effect on soft soils and improving on the volume changes that impair the overall long-term performance and service of the infrastructures constructed with these materials.

4   Conclusion

Solid waste recycling and reuse by direct and controlled combustion in a SWI-NaOH-OCEM to achieve a more environmental friendly, environmental efficient, and sustainable civil engineering infrastructure and a low or zero carbon emission into our planet has been reviewed. This review work among other things has developed a model for effective solid waste combustion, entrapping the oxides of carbon (CO_2 and CO) released hitherto, and the release of environmental friendly residue of baking soda ($NaHCO_3$), soda ash (Na_2CO_3) and hydrogen gas (H_2). This research also has shown the efficiency of using these ecofriendly materials in soil stabilization to improve the engineering properties of soft clay and expansive soils.

Conflict of Interest. The authors declare no conflict of interest in this work.

References

Abdel-Gawwadm, H.A., Abo-El-Enein, S.A.: A novel method to produce dry polymer cement powder. HBRC J. **12**, 13–24 (2016). https://doi.org/10.1016/j.hbrcj.2014.06.0018

Akbari, H., Mensah-Biney, R., Simms, J.: Production of geopolymer binder from coal fly ash to make cement-less concrete. In: World of Coal Ash (WOCA) Conference in Nasvhille, TN, 5–7 May 2015

Amadi, A.A., Eberemu, A.O.: Delineation of compaction criteria for acceptable hydraulic conductivity of lateritic soil-bentonite mixtures designed as landfill liners. Environ. Earth Sci. **67**, 999–1006 (2012). https://doi.org/10.1007/s12665-012-1544-z

American Standard for Testing and Materials (ASTM) C618, 1978 Specification for Pozzolanas. ASTM International, Philadelphia, USA

Eberemu, A.O., Amadi, A.A., Edeh, J.E.: Diffusion of municipal waste contaminants in compacted lateritic soil treated with bagasse ash. Environ. Earth Sci. (2012). https://doi.org/10.1007/s12665-012-2168-z

Eberemu, A.O., Afolayan, J.O., Abubakar, I., Osinubi, K.J.: Reliability evaluation of compacted lateritic soil treated with bagasse ash as material for waste land fill barrier. In: Geo-Congress 2014 Technical Papers, GSP, vol. 234, pp. 911–920 (2014)

Van Bui, D., Onyelowe, K.C.: Adsorbed complex and laboratory geotechnics of Quarry Dust (QD) stabilized lateritic soils. Environ. Technol. Innov. **10**, 355–368 (2018). https://doi.org/10.1016/j.eti.2018.04.005

Bui Van, D., Onyelowe, K.C., Nguyen Van, M.: Capillary rise, suction (absorption) and the strength development of HBM treated with QD base geopolymer. Int. J. Pavement Res. Technol. (2018). https://doi.org/10.1016/j.ijprt.2018.04.003

Ezugwu, C.N.: New approaches to solid waste management. In: Proceedings of the World Congress on Engineering and Computer Science WCECS 2015, vol. 11, San Francisco, USA, 21–23 October 2015

Eberemu, A.O.: Evaluation of bagasse ash treated lateritic soil as a potential barrier material in waste containment application. Acta Geotech. (2013). https://doi.org/10.1007/s11440-012-0204-5

Eberemu, A.O., Osinubi, K.J.: Comparative study of soil water characteristic curves of compacted bagasse ash treated lateritic soil. In: 6th International Congress on Environmental Geotechnics, pp. 1378–1383, New Delhi, India (2010)

Eberemu, A.O., Tukka, D.D., Osinubi, K.J.: The potential use of rice husk ash in the stabilization and solidification of lateritic soil contaminated with tannery effluent. In: ASCE Geo-Congress 2014 Technical Papers, GSP, vol. 234, pp. 2263–2272 (2014)

Ezugwu, C.N.: Ten segments of a comprehensive and cost-effective solid waste management system. The Nnewi Engineer. A Quarterly Newsletter of the Nig. Soci. of Engineers, Nnewi Branch, pp. 14–16, June 2015

Ezugwu, C.N., Uneke, L.A., Akpan, P.P.: Rice husk ash-an alternative to gypsum in POP board. Int. J. Eng. Sci. Math. **4**(4), 24–35 (2015)

Hamidi, R.M., Man, Z., Azizli, K.A.: Concentration of NaOH and the effect on the properties of fly ash based geopolymer. In: 4th International Conference of Process Engineering and Advanced Materials, Procedia Engineering, vol. 148, pp. 189–193 (2016). http://dx.doi.org/10.1016/j.proeng.2016.06.568

Hariz, Z., Mohd-MustafaAl-Bakri, A., Kamarudin, H., Nurliyana, A., Ridho, B.: Review of various types of geopolymer materials with the environmental impact assessment. In: MATEC Web of Conferences, vol. 97, p. 01021 (2017). http://dx.doi.org/10.1051/matecconf/20179701021

Ikeagwuani, C.C., Nwonu, D.C.: Emerging trends in expansive soil stabilization; a review. J. Rock Mech. Geotech. Eng. (2019). https://doi.org/10.1016/j.jrmge.2018.08.013

Ikeagwuani, C.C., Obeta, I.N.: Stabilization of black cotton soil subgrade using sawdust ash and lime. Soils Found. (2019). https://doi.org/10.1016/j.sandf.2018.10.004

Onyelowe, K.C., Bui Van, D., Nguyen Van, M., Ezugwu, C., Amhadi, T., Sosa, F., Wu, W., Ta Duc, T., Orji, F., Alaneme, G.: Experimental assessment of subgrade stiffness of lateritic soils treated with crushed waste plastics and ceramics for pavement foundation. Int. J. Low-Carbon Technol. 1–18 (2019). https://doi.org/10.1093/ijlct/ctz015

Onyelowe, K.C., Van, D.B., van Nguyen, M., Ugwuanyi, H.: Effect of ceramic waste derivatives on the volume change behavior of soft soils for moisture bound transport geotechnics. Electron. J. Geotech. Eng. **23**(04), 821–834 (2018)

Osinubi, K.J., Oluremi, J.R., Eberemu, A.O., Ijimdiya, S.T.: Interaction of landfill leachate with compacted lateritic soil–waste wood ash mixture. In: Proceedings of the Institution of Civil Engineers, Waste and Resource Management (2017). https://doi.org/10.1680/jwarm.17.00012Paper1700012

Osinubi, K.J., Eberemu, A.O., Amadi, A.A.: Compatibility of compacted lateritic soil treated with bagasse ash and municipal solid waste leachate. Int. J. Environ. Waste Manage. **10**(4), 365–376 (2012)

Onyelowe, K.C., Ubachukwu, O.A.: Stabilization of olokoro-umuahia lateritic soil using Palm Bunch Ash (PBA) as admixture. Umudike J. Eng. Technol. (UJET) **1**(2), 67–77 (2015)

Onyelowe, K.C., Okafor, F.O.: Review of the synthesis of nano-sized ash from local waste for use as admixture or filler in engineering soil stabilization and concrete production. J. Environ. Nanotechnol. (JENT) **4**(4), 23–27 (2015)

Onyelowe, K.C.: Nanosized palm bunch ash (NPBA) stabilisation of lateritic soil for construction purposes. Int. J. Geotech. Eng. (2017a). http://dx.doi.org/10.1080/19386362.2017.1322797

Onyelowe, K.C.: Nanostructured waste paper ash stabilization of lateritic soils for pavement base construction purposes. Electron. J. Geotech. Eng. **22**(09), 3633–3647 (2017b). www.ejge.com

Onyelowe, K.C.: Solid wastes management (SWM) in Nigeria and their utilization in the environmental geotechnics as an entrepreneurial service innovation (ESI) for sustainable development. Int. J. Waste Resour. **7**, 282 (2017c). ISSN: 2252-5211

Onyelowe, K.C., Maduabuchi, M.N.: Palm bunch management and disposal as solid waste and the stabilization of olokoro lateritic soil for road construction purposes in Abia State, Nigeria. Int. J. Waste Resour. **7**(2) (2017). https://doi.org/10.4172/2252-5211.1000279

Onyelowe, K.C., Ekwe, N.P., Okafor, F.O., Onuoha, I.C., Maduabuchi, M.N., Eze, G.T.: Investigation of the stabilization potentials of nanosized-waste tyre ash (NWTA) as admixture with lateritic soil in Nigeria. Umudike J. Eng. Technol. (UJET) **3**(1), 26–35 (2017c). www. ujetmouau.com

Onyelowe, K.C., Bui Van, D.: Durability of nanostructured biomasses ash (NBA) stabilized expansive soils for pavement foundation. Int. J. Geotech. Eng. (2018a). https://doi.org/10. 1080/19386362.2017.1422909

Onyelowe, K.C., Bui Van, D.: Predicting subgrade stiffness of nanostructured palm bunch ash stabilized lateritic soil for transport geotechnics purposes. J. GeoEng. Taiwan Geotech. Soc. **13**(2), 59–67 (2018b). http://140.118.105.174/jge/article.php?v=20&i=72&volume= 13&issue=2

Onyelowe, K.C., Bui Van, D., Van, M.N.: Swelling potential, shrinkage and durability of cemented and uncemented lateritic soils treated with CWC base geopolymer. Int. J. Geotech. Eng. (2018). https://doi.org/10.1080/19386362.2018.1462606

Osinubi, K.J., Eberemu, A.O.: Unsaturated hydraulic conductivity of compacted lateritic soil treated with bagasse ash. In: ASCE GeoFlorida 2010: Advances in Analysis, Modeling & Design (GSP 199), pp. 357–369 (2010)

Osinubi, K.J., Eberemu, A.O.: Effect of desiccation on compacted lateritic soil treated with bagasse ash. In: Materials Society of Nigeria (MSN) Zaria Chapter Book of Proceedings, 4th Edn, pp. 1–10 (2008)

Vikas, S., Atul, A.I., Mehta Satyendranath, P.K., Tripathi, M.K.: Supplementary cementitious materials in construction - an attempt to reduce CO2 emission. J. Environ. Nanotechnol. **7**(2), 31–36 (2018). https://doi.org/10.13074/jent.2018.06.182306

Treatment of Clayey Soils with Steel Furnace Slag and Lime for Road Construction in the South West of Iran

Ebrahim Asghari-Kaljahi[✉], Zahra Hosseinzadeh,
and Hadiseh Mansouri

Department of Earth Sciences, University of Tabriz, Tabriz, Iran
{e-asghari, h.mansouri}@tabrizu.ac.ir,
zahrahoseinzade024@gmail.com

Abstract. The fine-grained soils of Arvand free zone in the south west of Iran contains more than 95% fine grained particles. These soils cannot be considered as a proper material for earth works due to problems caused by clay soils such as high expansive, volumetric shrinkage, high settlement under loading and high moisture absorption. Meanwhile, there are no coarse-grained soil resources for using in the earth works around in this area. Khuzestan steel plant makes enormous amount of steel furnace slag as waste product and is available for any use. This study evaluates the treatment of fine grained soil by adding steel furnace slag and lime. For this purpose, soil was mixed with 10, 20 and 30% slag and 2, 4 and 6% hydrated lime. After curing the mixed soil, the change in soil characteristics were tested through Atterberg limits, compaction, unconfined compressive strength (UCS) and CBR tests. The test results showed that the soil plasticity and optimum moisture content decrease and maximum dry density, UCS and CBR increase by increasing slag and lime. The UCS of the soil is 147 kPa in the maximum density and is reaching to 267, 417 and 456 kPa by adding 30% slag and 2, 4 and 6% lime, respectively. The soaked CBR tests indicated that adding 20% slag and 4% lime provide a CBR more than 30%. This mixture is suitable from technical and economical views and is recommended for soil treatment of the study area and using in earth works, such road construction.

Keywords: Arvand free zone · Steel slag · Lime · Soil treatment

1 Introduction

The fine-grained soils have various problems like low compressive and shear strength, instability, expansive nature, volumetric shrinkage, cracking due to desiccation, high settlement under loading and low durability against atmospheric factors such as wetting and drying and/or thawing and freezing periods (Makarchian and Naderi 2010; Tangri and Gagandeep 2018). These problems restrict the civil engineers to use these kind of soils for earth works or built large construction projects on them (Tangri and Gagandeep 2018). Engineering properties of these soils can be improved by using certain additives. Recently, improvement of soft problematic soils by using industrial by-

© Springer Nature Switzerland AG 2020
H. Ameen et al. (Eds.): GeoMEast 2019, SUCI, pp. 68–78, 2020.
https://doi.org/10.1007/978-3-030-34199-2_5

products (e.g. slag or fly ash) has been considered to be an economically and environmentally relabel solution (James et al. 2008; Nidzam and Kinuthia 2010; Wilkinson et al. 2010 and Gonawala et al. 2018). Steel slag, a by-product of steel making, is mostly composed of a complex solution of silicates and oxides. It is a recycled material that can be used as an aggregate replacement in many civil applications including in road base, concrete mixes, asphalt concrete mixes and soil stabilization (Aldeeky and Al Hattamleh 2017). Due to lack of proper aggregates in many parts of Iran, including south of Khuzestan, the civil projects are more difficult and costly to build. The use of steel slag as a low cost soil modifier in these areas can decrease construction costs and help to maintain of non-renewable resources such as sand and gravel.

Previous researchers were evaluating the effectiveness of steel slag to enhance mechanical properties of problematic soils. Akinwumi (2014) showed that the addition of steel slag to lateritic soil results in a decrease in Atterberg limits and optimum moisture content and an increase in maximum dry density. They also showed that unsoaked and soaked California Bearing Ratio (CBR), the unconfined compressive strength (UCS) and permeability of the soil increase with increasing slag content. Experimental work by Zumrawi and Babikir (2017) indicate that addition of 30% slag to the soil reduces the free swell index of the soil by about 55%. As shown by Shalabi et al. (2017) the cohesion of the clay soil decreases and its internal friction angle increases with increasing slag content. The slag can also control swell associated with sulphate bearing soils (Makarchian and Naderi 2010; Celik and Nalbantoglu 2013). The abnormal plasticity and swell potential occurs in the soil due to presence of sulphate. The laboratory test results conducted by Celik and Nalbantoglu (2013) reveal that at 10000 ppm sulphate concentration, the swell potential of the lime stabilized soil is three times larger than the natural soil. They show that the swell potential of this soil reduces about 87.5% by adding 6% granulated steel slag. Gonawala et al. (2018) searched on the stabilization of expansive soil by adding Corex slag and lime. Corex slag was mixed with the soil in the range of 10 to 30% with an increment of 5% and lime was added by 2% and 4% in the mix by the dry weight of soil. Admixing of all these stabilizers improves soaked CBR and UCS values. The addition of 25% Corex slag with 4% lime in the soil gives an optimum mixture. The UCS and CBR values of the soil increased from 0.24 MPa to 1.09 MPa and from 1.86 to 53.5% for the optimum combination.

The main aim of the current study is to evaluate the use of steel slag and lime for improving the mechanical properties of the fine-grained soil which covers the most parts of the Arvand free zone in Khorramshahr area, south west of Iran. This soil has a low bearing capacity and is not proper for using in road subgrade or subbase, foundation and other construction projects. It should be noted the coarse aggregate resources are located at the large distance from this area and subsequently using of them for the civil projects is so costly. However, large amounts of steel slag are daily producing in Khuzestan steel plant processes that are dumped in open areas. The study considers the effect of the steel slag and lime on Atterberg limits, compaction properties, UCS and CBR of the treated soil.

2 Materials and Tests

The clayey soil used in this research was taken from the Arvand free zone located in Khuzestan province, south west of Iran (Fig. 1). To characterize the soil sample, grain size distribution (ASTM D7928), Atterberg limits (ASTM D4318) and modified compaction (ASTM D1557) tests were performed and their results are shown in Table 1. Based on the particle-size analysis and plasticity limits, the soil is classified as CL according to USCS with more than 95% fine material (Fig. 2). The plasticity index of the soil is about 27%. The result of modified compaction test showed that the optimum moisture content and the maximum dry density of the soil are 22.5% and 16.1 kN/m^3, respectively. The soil contains 4 to 8% of the water soluble materials. The soil sulphate concentration of this area is relatively high, between 0.5 and more than 4% (Mandro Consulting Engineers 2014).

Fig. 1. Location of the study area

The steel furnace slag aggregate was obtained from Khuzestan (Ahwaz) steel plant. It has the specific gravity about 34 kN/m^3 and water absorption content about 0.7%. The grain size of slag is various, from powder size to more than 40 cm (Fig. 3), but is used passing through sieve No. 10 (grain size less than 25 mm). The Los Angeles abrasion value (mass loss) of the slag was tested as 4.5% after 100 revolutions and 17% after 500 revolutions; therefore, the slag grains are so resistant. The hydrated lime was provided from Haftgol factory as a white powder.

Table 1. Physical properties of the study soil

Grain size			Atterberg limits			Modified compaction	
Clay (%)	Silt (%)	Sand (%)	LL (%)	PL (%)	PI (%)	Maximum dry density (kN/m³)	Optimum moisture content (%)
40	55	5	49	22	27	16.1	22.5

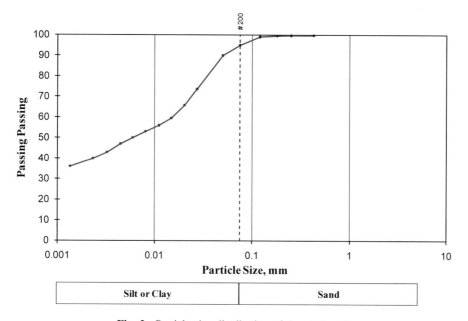

Fig. 2. Particle size distribution of the study soil

The steel slag with amounts of 10, 20 and 30% and hydrated lime with amounts of 2, 4 and 6% by dry weight of the soil were added and mixed with dry soil. The Atterberg limits, UCS and CBR tests were performed on the soil-slag and lime mixtures samples. The optimum moisture content and maximum dry density (is determined by the modified compaction test) were used to prepare the samples.

For Atterberg tests, each soil-additive mixture was passed through sieve No. 40 and liquid limit (LL) and plastic limit (PL) were determined after 28 days curing. The UCS tests (ASTM D2166) were performed on the soil-additives mixtures passing sieve No. 8 (2.4 mm). The samples were molded in a split steel tube of 50 mm diameter and 100 mm height and compressed to the desired compaction. After 28 days curing, the samples were tested by UCS apparatus. The CBR tests (ASTM 1883) were performed in two conditions: one test was done immediately after sample preparation (named unsoaked CBR test), and the other sample after soaking in water for 4 days (named soaked CBR test). When the soil mixture contained the lime as an additive, the CBR test was performed after curing for 28 days.

Fig. 3. An image of the used steel slag aggregate

3   Results and Discussion

Atterberg Limits

Figure 4 shows the effect of slag and lime content on the liquid limit, plastic limit and plasticity index. This figure shows that the Atterberg limits decrease with increasing slag and lime content. For example, the plasticity index is 24% for 10% slag and 2% lime and decreases by 12% for 30% slag and 6% lime. As concluded by Aldeeky and Al Hattamleh (2017), the reduction in the soil plasticity with increasing slag content can be attributed to increase in the sand and silt size particles in the soil mixture. The decrease in the Atterberg limits with increasing lime content can be due to the replacement of lower valance cations in the soil by Ca^{2+} and Mg^{2+} in the lime powder

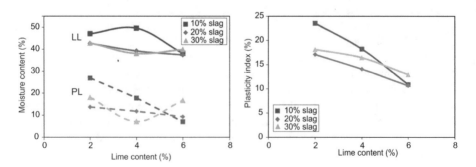

Fig. 4. Variation of Atterberg limits with steel slag and lime contents

and subsequently the reduction of diffused water layer around the clay particles (Akinwumi 2014). Figure 4 shows that for the slag content more than 20% and lime content more than 4%, the reduction in Atterberg limits, especially in the liquid limit is negligible.

Compaction Test

The modified compaction tests were performed on the soil mixtures with 10, 20 and 30% slag according to ASTM D1557. Figure 5 shows the values of dry density and optimum moisture content obtained for different slag contents. This figure shows the dry density increases and the optimum moisture content decreases with increasing slag content. At 10% slag content, the dry density and optimum moisture content were obtained as 16.7 kN/m^3 and 20.5%, respectively. These values respectively reached to 19.5 kN/m^3 and 14.5% for the 30% slag content.

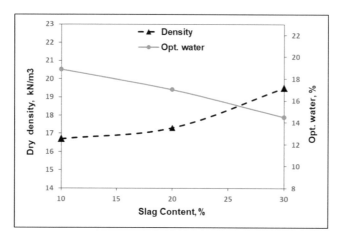

Fig. 5. Variation of dry density and optimum water content with steel slag content in the modified compaction test

Uniaxial Compressive Strength

The mean UCS values of the soil for various slag and lime contents are presented in the Table 2. It is obvious that the UCS of the soil increases with increasing slag and lime contents. For example, it is 147 kPa for the soil in the maximum density condition and increases to 456 kPa for the soil mixed with 30% slag and 6% lime. As stated before, the increasing of UCS can be due to increasing sand and silt size particles in the soil and pozzolanic compounds which are produced during the reaction between soil and lime (Tangri and Gagandeep 2018). The stress-strain curves of some soil-additives mixtures at constant slag and lime content are shown in Fig. 6. These curves show that the 20% slag and 4% lime can be the proper contents for stabilization of the study soil. As seen, the increase in UCS value of the soil mixtures is not significant for the slag and lime contents more than 20% and 4%, respectively. Figure 7 shows some samples after the UCS test. As seen, the failure mode is relatively similar in all samples;

however, the failure surface angle slightly decreases as slag content increases. In fact, the failure mode changes from compressive failure to shear failure with increasing slag and lime contents, which can be due to change in the soil stiffness.

Table 2. The mean values of UCS for the various soil-slag and lime mixtures

Slag content, %	Lime content, %	UCS, kPa
0	0	147
10	2	226
	4	343
	6	428
20	2	265
	4	408
	6	443
30	2	267
	4	411
	6	456

California Bearing Ratio (CBR)

CBR samples were prepared with various contents of steel slag and lime. The un-soaked and soaked CBR values of soil were about 19% and 7%, respectively. CBR samples with slag contents 10, 20 and 30% and also lime contents 2, 4 and 6% were prepared and cured for 28 days. Table 3 shows the CBR values obtained for the soil with different slag and lime contents. This table also shows the swell potential of each soil mixture after soaking. As seen, the effect of slag in reducing the swell percent is not significant and the swell percent has increased after adding 20 and 30% slag to the soil. In contrast, the lime has a considerable effect on the swell percent reduction, such that the swell percent decreases from 2% to lower than 0.2% by increasing the lime content.

The variation in the soaked and un-soaked CBR values with the slag and lime contents is shown in Fig. 8. The CBR values increase with increasing slag and lime contents, that is similar to findings of Shalabi et al. (2017). As seen in Fig. 8a, at more than 20% slag content, the effect of lime on increasing the un-soaked CBR value becomes insignificant. Figure 8b shows that the effect of lime on increasing the soaked CBR value is higher than the effect of slag. For example, at 10% slag and 2% lime, the soaked CBR value was obtained as 16.4%, by keeping the lime content and increasing slag content to 20%, the soaked CBR value increases by 19%, however it increases by 34% for the 10% slag and 4% lime. As seen in Fig. 8b, for the soil mixtures containing 2% lime, the slag content cannot considerably enhance the soaked CBR value even in the percentage of 30% (Fig. 8b).

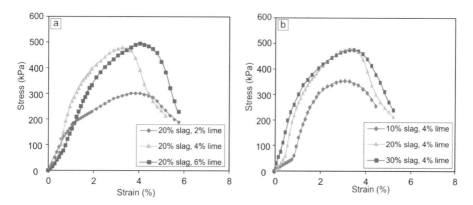

Fig. 6. Some stress-strain curves of the soil mixtures at constant slag content (a) and constant lime content (b)

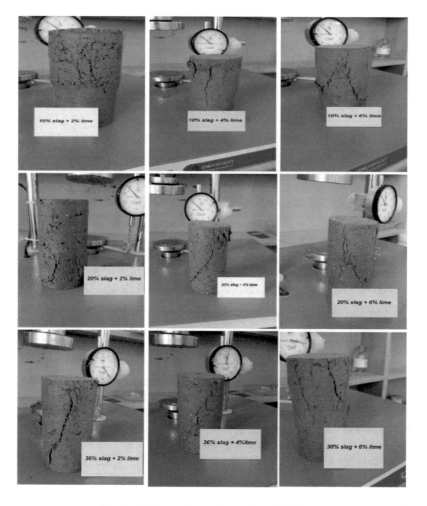

Fig. 7. Failure pattern of samples at UCS tests

Table 3. The results of CBR tests for different soil- slag and lime mixtures

Slag content (%)	Lime content (%)	Soaked CBR (%)	Unsoaked CBR (%)	Swelling (%)
10	2	16	28	1.0
	4	33	46	0.2
	6	44	55	0.2
20	2	20	34	0.8
	4	41	58	0.4
	6	51	63	0.4
30	2	24	62	1.5
	4	50	69	0.5
	6	55	75	0.5

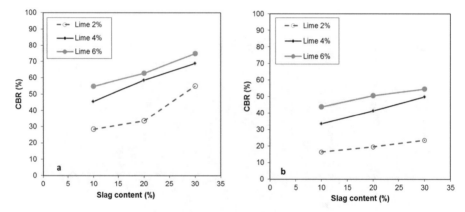

Fig. 8. The CBR for different mixtures of soil + slag + lime in unsoaked (a), and soaked (b) conditions

The results of soaked CBR suggest that 20% slag and 4% lime can be the optimum amounts of additives for achieving to CBR value 30%, which is suggested for the road subbase layer by Iranian road specification code (Plan and Budget Organization of Iran 2013).

4   Conclusions

The fine-grained soils of Arvand free zone contains more than 95% clay and silt and is not a proper material for using in the earth works for road construction. This study aimed to improve this soil by adding steel slag and hydrated lime. Since the sulphate concentration was relatively high in the soil of this area of Iran, the use of lime along with the slag was suggested for the soil improvement. The following results can be drowning from the laboratory tests:

- The liquid limit, plastic limit and plasticity index decreased as slag and lime contents increased in the soil.
- By increasing the amount of slag, the optimum moisture content of the soil decreased and its dry density increased.
- The UCS value of the soil was 147 kPa in the maximum density. It increased with increasing slag and lime content, for example the UCS value of the soil mixed with slag 30% and lime 6% increased to 456 kPa.
- The soaked and un-soaked CBR values increased with increasing slag and lime contents. The results of CBR tests suggested that 20% of slag and 4% of lime in the soil could be the optimum contents for improving the soil characteristics and reaching to the CBR value needed for road subbase layer.

Acknowledgments. Thanks to the Jahan Ara Arvand Steel Co. and P. O. Rahvar Consulting Engineers Co. for providing information and assistance in sampling.

References

Akinwumi, I.: Soil modification by the application of steel slag. Periodica Polytech. Civil Eng. (2014). https://doi.org/10.3311/ppci.7239

Aldeeky, H., Al Hattamleh, O.: Experimental study on the utilization of fine steel slag on stabilizing high plastic subgrade soil. Adv. Civil Eng. (2017). https://doi.org/10.1155/2017/9230279

Tangri, A., Gagandeep: Effect of blast furnace slag on various properties of clayey soil: a review. Int. J. Sci. Res. Dev. **6**, 2321–0613 (2018)

ASTM: Annual book of standards. American Society of Testing and Materials, West Conshohocken (2018)

Celik, E., Nalbantoglu, Z.: Effects of ground granulated blastfurnace slag (GGBS) on the swelling properties of lime-stabilized sulfate-bearing soils. Eng. Geol. (2013). https://doi.org/10.1016/j.enggeo.2013.05.016

Gonawala, R.J., Kumar, R., Chauhan, K.A.: Stabilization of expansive soil with Corex slag and lime for road subgrade. In: McCartney, J.S., Hoyos, L.R. (eds.) GeoMEast 2018, SUCI, pp. 1–14 (2018). https://doi.org/10.1007/978-3-030-01914-3_1

James, R., Kamruzzaman, A., Haque, A., Wilkinson, A.: Behaviour of lime–slag-treated clay. Proc. Inst. Civil Eng. Ground Improv. (2008). https://doi.org/10.1680/grim.2008.161.4.207

Makarchian, M., Naderi, H.: Effect of moisture on CBR strength of soil stabilized by lime and ground granulated blast furnace slag (GGBS) in the precence of sulphate. In: 5th National Congress of Civil Engineering, Ferdowsi University of Mashhad (2010)

Mandro Consulting Engineers Co.: Geotechnical report of Arvand complex project (2014)

Nidzam, R., Kinuthia, J.M.: Sustainable soil stabilisation with blastfurnace slag–a review. Proc. Inst. Civil Eng. Const. Mater. (2010). https://doi.org/10.1680/coma.2010.163.3.157

Plan and Budget Organization of Iran: Road general technical specification, Department of Technical Affairs, Code No. 101 (2013)

Shalabi, F.I., Asi, I.M., Qasrawi, H.Y.: Effect of by-product steel slag on the engineering properties of clay soils. J. King Saud Univ. Eng. Sci. (2017). https://doi.org/10.1016/j.jksues.2016.07.004

Wilkinson, A., Haque, A., Kodikara, J.: Stabilisation of clayey soils with industrial by-products: part A. Proc. Inst. Civil Eng. Ground Improv. (2010). https://doi.org/10.1680/grim.2010.163.3.149

Zumrawi, M.M., Babikir, A.A.-A.A.: Laboratory study of steel slag used or stabilizing expansive soil. Univ. Khartoum Eng. J. **4**, 1–6 (2017)

"A Literature Review on Solid Waste Management: Characteristics, Techniques, Environmental Impacts and Health Effects in Aligarh City", Uttar Pradesh, India"

Harit Priyadarshi[1]([✉]), Sarv Priya[2], Ashish Jain[1],
and Shadab Khursheed[3]

[1] Department of Civil Engineering, Mangalayatan University,
Beswan, Aligarh 202145, Uttar Pradesh, India
gsiharit@rediffmail.com
[2] Department of Civil Engineering, KIET, Murad Nagar, Ghaziabad 201206,
Uttar Pradesh, India
[3] Department of Geology, Aligarh Muslim University,
Aligarh 202002, Uttar Pradesh, India

Abstract. India is known as one of the most heavily settled countries in the world. It appears to be the second country to have the highest number of residents. With the total population of about expected data 1.37 billion in 2019. The management of Municipal Solid Waste (MSW) in India has encountered problems. Each year, the population grew by 3–3.5%, as this factor arises, the rate of solid waste generation also rise up to 1.3% in Aligarh city, Uttar Pradesh a large number of ingenious factors like, rapid urbanization, rapid population density, rapid commercialization, uneven living standards and also enlargement of industrialization has created destructive consequences in terms of biodegradable and non-biodegradable waste generations which are estimated at about 415 tons per day.

This paper emphasizes the waste characteristics, techniques, adverse environmental impacts, health risks, poor waste management practices and also problems associated with the solid waste management system at the municipal level.

The findings from this study indicates failure of the existing facilities due to lack of concern, high volume of waste generation, deficient collection space, delayed sanctioning of new landfill sites and a number of open-dump sites which generate fires. The innuendos of the waste management practices in the city are discussed.

Keywords: Sources of M.S.W · Component of M.S.W · Health risks and sustainable approaches

© Springer Nature Switzerland AG 2020
H. Ameen et al. (Eds.): GeoMEast 2019, SUCI, pp. 79–90, 2020.
https://doi.org/10.1007/978-3-030-34199-2_6

1 Introduction

"Let us keep our city clean"

In recent years fast population growth, increase in urbanization and industrialization in India has created severe problems for solid waste management in cities. The increased level of consumption characteristics of the population of cities lead to generation of enormous quantities of solid waste material. The impacts of such pollution are felt both at local, as well as, at distances from sources. Domestic and industrial discharges lead to contamination of air, eutrophication with nutrient and toxic materials which in turn lead to degradation of air, land and affect flora and fauna badly. Since olden times municipal bodies remained responsible for keeping the roads clean, collect city garbage and to carry out its safe disposal. Most of the elected bodies of the Indian cities employ largest number of employees for the purpose of cleaning the city, but only 50–70% of the waste generated is collected by the staffkeeping aside the tendency of nonworking of the employees. Many estimates of solid waste generation are available but on the average it is projected that under Indian conditions the amount of waste generated per capita will rise at a rate of 1–1.33% annually (Shekdar 1999). So, at present if we follow this presumption the calculated per capita waste generation on daily basis is 583.36 g in 2016. At such a stage solid waste generation will have significant impact in terms of land required for disposal of waste as well as methane emission. Such a large quantity of solid waste requires well managed system of collection, transportation and disposal. It is required that we have proper knowledge about the nature of waste material, its collection and disposal along with recycling and energy generation potential. The traditional routine approach to solid waste management is normally municipal bodies handle all aspects of collection, transport and disposal and this has emerged as a reality of mixed success all over the world in advanced or developing cities. The search for more efficient and economical solid waste collection agenda in most of the urban areas has taken shape adopting several directions towards better partnership with communities along with private sector combining adequate economic policies, e.g., recycling credits by paying the recycler, land-fill disposal levies at land-fill sites designed to minimize the quantity of waste being land-filled and product charges like packing tax to disallow over-packaging. Cities have a wide variety of arrangement under their control to lessen environmental burdens. Legal approach and restrictions on the quantity of pollutants a factory can discharge of minimum air and water quality standards are being particularly proved effective in monitoring pollution in many parts of the globe. The efficiency depends mainly on good enforcement capacities and proper monitoring procedures where urban growth pressures and pollution issues are far greater. The present scenario of solid waste management in Aligarh city shown in Table 1.

Table 1. The present scenario of solid waste management in Aligarh city.

Functional element	Detail
Segregation of storage at source	Generally absent, waste is thrown on streets
Primary collection	Does not exist, waste is deposited on the streets and picked up through sweeping
Waste storage deposits	Very unscientific, waste is stored on open sites/ Masonry enclosures. A few containers are however, is use
Transportation	Manual loading is open trucks/ Partly dumper placers
Frequency of removal	Irregular/Alternate day/ Once in three days/ Once in a week
Processing	No processing is carried out except A to Z municipal power plant
Disposal	Unauthorized dumping in open space

(Source solid waste management, NEERI vol. 35, 2004).

2 Study Area

The Aligarh is an ancient city in the north Indian state of Uttar Pradesh is situated in the middle of doab-the land between the Ganga and Yamuna rivers, at a distance of 130 km Southeast of Delhi on the Delhi- Howrah rail route and the Grand Trunk road. Aligarh lies between latitude 27° 54′ and 28° north and Longitude is 78° and 78° 5′ east, shown in Fig. 1. The Aligarh city is spread over an area of about 36.7 km². The area lies between the Karwan River in the west and the Senger River in the east and is a part of central Ganga basin. It is the administrative headquarter of Aligarh division. Aligarh is mostly known as a university city where the famous Aligarh Muslim University is located. The Aligarh city is an important centre of lock smithy and brassware

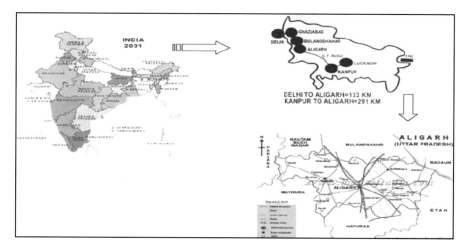

Fig. 1. Location of the study area (Aligarh city), Uttar Pradesh, India.

manufacturing. There are a total of 5506 industrial units in Aligarh city, of these; there are 3500 small scale industries, 2000 medium scale 6 large industries.

3 Objectives

The purpose of this study is to assess the current practices and state of solid waste management systems (SWMS) in one medium-sized Indian town, identifying main issues and problems to its ineffectiveness, inefficiency and to gain some suggestions and recommendations to improve the SWM infrastructure and practices in such Indian towns. Municipal solid material generation and their disposal is a major and critical issue in almost all municipal cities of India. It can harm local environment, as well as, pollute underground potable water. It may also become responsible for dissemination of various diseases in urban areas and its peripheries. Present investigation has been planned to include the target of municipal solid waste management by reducing the quantity of routine production of waste and proper disposal of waste along with recovery of materials and energy from solid waste. All such practices do not have much requirement of any kind of specific raw material and energy inputs technological processes. The proposed investigations have the following objectives of the study for proper management of municipal solid waste in Aligarh city:

(1) To estimate quantum and prevailing treatment practices of municipal solid waste in study area.
(2) To analyze various properties and environmental impact of municipal solid waste in study area.
(3) To develop strategies for suitable collection, segregation, recycling treatment methods for municipal solid waste in study area.
(4) To assess the use of municipal solid waste through eco-friendly methods and application of municipal solid waste for different uses in study area.
(5) Develop strategy for mitigation of carbon-di-oxide potential through municipal solid waste disposal methods with the updated data in study area.

4 Scope of the Work

1. The study involves door-to-door survey in the residential area of the Aligarh city.
2. The primary data collection including: -

- Generation of the solid waste in the Aligarh city.
- Source of the solid waste in the Aligarh city.
- Quantity of the waste generated in the Aligarh city.
- Health Status of the city in the Aligarh city.
- About the disposal methodology of the waste in the Aligarh city.
- Help in the comparison of the previous data.

4.1 Methodology

A medium size city, Aligarh, having a population estimate 1.36 million in 2019, and situated 130 km from the capital city of India, Delhi, was selected for this study. An extensive literature review was conducted to establish a theoretical framework. Field visits were conducted to collect the primary data and to understand the solid waste management of the city.

5 Effects on Poor Waste Management

Health issue is the major problem in India as many of the disease came from the pollutions made by them Health issue arise due to poor waste management in study area for example is malnutrition, especially the children which is the condition that develops when the body does not get the right amount of the vitamins, minerals, and other nutrients it needs to maintain healthy tissues and organ function. (Medical dictionary, 2012) Furthermore, health issue such as dengue, fever, Hepatitis, tuberculosis, malaria, pneumonia, and also poor sanitation due to poor waste management. Due to poor waste management by the authorities, availability of clean and safe water is minimized because of people threw rubbish at the river and the quality of living will decrease.

6 Technique Action

6.1 Municipal Solid Wastes Collection

State government should enforce new strategies which prohibit littering of municipal solid waste in cities towns and urban areas. The following steps shall be taken by the municipal authority.

1. **Organizing house-to-house collection of municipal solid wastes:** Through any of the methods, for example community bin collection (central bin), house-to-house collection, collection on regular pre-informed timings and scheduling by using bell ringing of musical vehicle (without exceeding permissible noise levels), Planning a systematic way and united effort for collection of waste from poverty areas or localities including hotels, restaurants, office complexes and commercial areas. Bio-medical wastes and industrial wastes shall not be combined with municipal solid wastes and such wastes should follow the rules separately specified for the purpose. Horticultural and construction or demolition wastes or debris shall be separately collected and disposed off following proper norms. Similarly, wastes generated at dairies shall be regulated in accordance with the State laws. Stray animals such as dogs and cats shall not be allowed to move around waste storage facilities or at any other place in the city or town and shall be managed in accordance with the State laws. The municipal authority shall notify waste collection schedule in neighborhoods.

2. **Segregation of municipal solid wastes:** Segregation materials should be done by municipal authority by promote recycling and reused waste by create or organized an awareness programs and campaign. The municipal authority shall take in charge phased programs to ensure community participates in waste segregation programmed. For this purpose, the municipal authorities shall arrange regular meetings at quarterly intervals with representatives of local resident welfare associations and non-governmental organizations.

3. **Storage of municipal solid wastes:** Municipal authorities shall establish and maintain storage facilities in such a manner as they do not create unhygienic and in sanitary conditions around it. There is some example criteria shall be taken to establishing and maintaining storage facilities.

    The quantities of waste generation should be counted in order to create enough storage facilities in a given area and the population densities. Furthermore, a storage facility shall be so placed that it is accessible to user. Storage facilities to be set up by municipal authorities or any other agency shall be so designed that wastes stored are not exposed to open atmosphere and shall be aesthetically acceptable and user-friendly.

4. **Transportation of municipal solid wastes:** Vehicles used for transportation of wastes shall be covered. Waste should not be visible to public, nor exposed to open environment preventing their scattering and unpleasant smell. The following criteria shall be met is the storage facilities set up by municipal authorities shall be daily attended for clearing of wastes. The bins or containers wherever placed shall be cleaned before they start overflowing.

5. **Processing of municipal solid wastes:** To minimize burden on landfill the municipal authorities shall adopt suitable technology or combination techniques to process the municipal solid waste. The biodegradable wastes shall be processed by composting, vermicomposting, anaerobic digestion or any other appropriate biological processing for stabilization of solid waste. Mixed waste containing recoverable resources shall follow the route of recycling. Incineration with or without energy recovery including pelletisation can also be used for processing wastes in specific cases. Municipal authority or the operator of a facility wishing to use other state-of-the-art technologies shall approach the Central Pollution Control Board to get the standards laid down before applying for grant of authorization.

7   Health Effects Due to Solid Waste

Over 3.6% annual growth in urban population and the rapid pace of urbanization, the situation is becoming more and more critical with the passage of time. Lack of financial resources, institutional weakness, improper choice of technology, and lack of support from public, towards Solid Waste Management (SWM) has made this service far from satisfactory. Waste generation ranges from 200 Gms to 500 Gms per capita per day in cities ranging from 1 lakh to over 50 lakhs population, as shown in Table 2.

As per above table per capita waste generation in Aligarh should be in the range of 270 gms (approximately). The larger the city, the higher is the per capita waste

Table 2. Waste generation per capita

Population range (in lakhs)	Avg. per capita waste generation gms/capita/day
1–5	210
5–10	250
10–20	270
20–50	350
50 & above	500

(Source www.jnnurm.com/india)

generation rate. The total waste generation in urban areas in the country is exceeded 39 million tons a year-by-year 2001, and estimated at 62 million tons a year-by year 2025. It is estimated that about 80,000 metric tons of solid waste is generated everyday in the urban centers of India. At present about 60% of the generated solid wastes is collected and unscientifically disposed off. The uncollected solid wastes remain in and around the locality or find its way to open drain, water bodies, etc.

The above information suggest that environmental pollution (specifically air pollution, water pollution and pollution due to waste) is a major health risks to humans, which is to be tackled on a priority basis. It is also essential to prepare compressive national health profile database on health effects due to pollution with respect to urban cities.

Solid waste not only affects the person living nearby it but also it used to affect other too. Actually solid waste due to the formation of the leachate and gases affect the water under the landfill site and also the air around it. Due to the formation of methane and other gases at the landfill site the atmosphere get distorted suddenly and harm the surrounding environment.

8 Discussion

Despite all efforts being made by the local municipality within their limited resources, the solid waste management situation in Aligarh is still not adequate. The waste is being dumped on low lying or open areas in the outskirts of the city without engineering and scientific methods. This situation of SWMS can be compared with other Indian towns of similar size. Management of municipal solid waste in the Aligarh city is far from satisfactory. There are problems in the solid waste management practices prevailing in the study area at every level, such as collection, transportation, processing and disposal. Mismanagement of solid waste is a matter of serious concern for public health and environment.

The generation of the organic manure to promote derivation manure from waste to reduce the quantity of waste going to landfill sites and also to help agricultural production, shown in Table 4. Aligarh Municipal corporation (AMC) tied up with A 2 Z for processing of 250 MT of garbage daily, out of which 42.5 MT of composed per day are produced by microbial compositing. AMC has further extended the waste processing facility to take one step ahead for sustainable waste management and to reduce the load on landfill sites. The following Field photographs (Figs. 2, 3, 4, 5, 6, 7, 8 and 9) are collected from study area.

9   Field Photographs

See Figures 10 and 11 and Table 3.

Fig. 2. Shows drains at Jamalpur, Aligarh.

Fig. 3. Shows cleaning drains at Goolar road, Aligarh.

Fig. 4. Shows drains at Jamalpur, Aligarh.

Fig. 5. Shows garbage FM tower near, Maheshpur.

Fig. 6. Goolar road Aligarh.

Fig. 7. Shows Trommel at A2Z Aligarh.

Fig. 8. Shows MSW spillage on the road (Bara Dwari).

Fig. 9. Pratibha Colony, infront of Nigam Workshop, Aligarh, U.P.

Table 3. Shows composition of waste

Composition of waste:	Percentage (%)
Organic	50
Recyclable	15
Silt & sand	15
Construction	10
Other waste	10

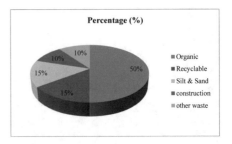

Fig. 10. Graphical representations on composition of waste in study area.

Table 4. Shows composition of manure.

Composition of manure:	Percentage (%)
Moisture	20
Particle size	91
Total organic matter	17.2
Total organic carbon	10
Density	0.95
Total nitrogen	0.48%
C:N ratio	10.65
Phosphores	0.4
Potash	0.4
Color	Black or brown
Odor	Odorless

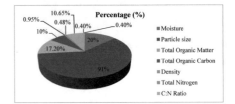

Fig. 11. Graphical representations on composition of manure in study area.

10   Conclusion

Some suggestions would be helpful in improving SWM system in Aligarh city these are as follow:

1. Open waste storage sites and other unhygienic street bins should not be allowed.
2. The placement of waste receptacles should be correct.
3. Door to door collection of waste must be made mandatory that will allow minimum of waste on roads and streets.
4. Land filling must be done properly after consideration of all the aspects of present and future of the city and its health
5. Alternative and better options for proper waste disposal method must be adopted regularly based on the needs and situation of the area,
6. There must be total ban on stray animals who wander on the roads which include cows, bulls, dogs, goats, etc. and these animals must be regularly trapped without any political or community influence. It will solve many of the problems associated with waste disposal.

7. Segregation of household waste at the source would reduce the burden of solid waste significantly while at the same time improve the supply of composite serving the nutrient poor farmer near Delhi.
8. Proper maintenance of vehicles and other equipments.
9. Government should adopt 4R's (Reduce, Reuse, Recycle and Resource Recovery) principle.
10. Government should increase the number of composting and energy generation plant.

Acknowledgement. First of all I would like to thanks Hon'ble Prime minister Narendra Damodar Das Modi to continue the dream of Mahatma Gandhi Make in India mission "Mission for sanitary India. The authors acknowledge all the persons involved in Aligarh Municipal Corporation (AMC) for providing all the pertinent information.

References

Lavee, A., Vievek, : Is municipal solid waste recycling economically efficient. Environ. Manag. **40**, 926–943 (2009)

Asnani, P.U.: United States Asia Environmental Partnership Report, United States Agency for International Development, Centre for Environmental Planning and Technology, Ahmedabad (2004)

Bhoyar, R.V., Titus, S.K., Bhide, A.D., Khanna, P.: Municipal and industrial solid waste management in India. J. IAEM **23**, 53–64 (1996)

Central Pollution Control Board (CPCB): Management of Municipal Solid Waste. Ministry of Environment and Forests, New Delhi, India (2004)

Zhu, D., Asnani, P.U., Zurbrügg, C., Anapolsky, S., Mani, S.: Improving Municipal Solid Waste Management in India (2008)

Ravi, D.: Solid waste management issues and challenges in Asia, Asian Productivity Organization (2007)

Dayal, G.: Solid wastes: sources, implications and management. Indian J. Environ. Prot. **14**(9), 669–677 (1994)

Garg, S., Prasad, B.: Plastic waste generation and recycling in Chandigarh. Indian J. Environ. Prot. **23**(2), 121–125 (2003)

Gupta, S., Krishna, M., Prasad, R.K., Gupta, S., Kansal, A.: Solid waste management in India: options and opportunities. Resour. Conserv. Recycl. **24**, 137–154 (1998)

Priyadarshi, H., Rao, S., Khan, S.S., Vats, D.: "Mission for the sanitary India: a case study of Aligarh City" Uttar Pradesh, India. In: Towards Sustainable Cities in Asia and the Middle East, International Congress and Exhibition "Sustainable Civil Infrastructures: Innovative Infrastructure Geotechnology", pp. 47–62. Springer publication Book Series (2018)

Priyadarshi, H., Jain, A.: Municipal solid waste management study and strategy in Aligarh City, Uttar Pradesh India. Int. J. Eng. Sci. Invent. (IJESI) **7**(5), 29–40 (2018). Ver. III

Joseph, K.: Perspectives of solid waste management in India. In: International Symposium on the Technology and Management of the Treatment and Reuse of the Municipal Solid Waste (2002)

Khan, R.R.: Environmental management of municipal solid wastes. Indian J. Environ. Prot. **14**(1), 26–30 (1994)

Kumar, S., et al.: Assessment of the status of municipal solid waste management in metro cities, state capitals, class I cities, and class II towns in India: an insight. Waste Manage. **29**(2), 883–895 (2009)

Kumar, S., Mondal, A.N., Gaikwad, S.A., Devotta, S., Singh, R.N.: Qualitative assessment of methane emission inventory from municipal solid waste disposal sites: a case study. Atmos. Environ. **38**, 4921–4929 (2004)

NEERI: Strategy Paper on SWM in India, National Environmental Engineering Research Institute, Nagpur (1995)

Schubeler, P.: NEERI Report "Strategy Paper on Solid Waste Management in India", pp. 1–7 (1996)

Rajput, R., Prasad, G., Chopra, A.K.: Scenario of solid waste management in present Indian context. Caspian J. Environ. Sci. **7**(1), 45–53 (2009)

Rao, K.J., Shantaram, M.V.: Physical characteristics of urban solid wastes of Hyderabad. Indian J. Environ. Prot. **13**(10), 425–721 (1993)

Rathi, S.: Alternative approaches for better municipal solid waste management in Mumbai. India J. Waste Manage. **26**(10), 1192–1200 (2006)

Sharma, S., Shah, K.W.: Generation and disposal of solid waste in Hoshangabad. In: Book of Proceedings of the Second International Congress of Chemistry and Environment, Indore, India, pp. 749–751 (2005)

Shekdar, A.V.: Municipal solid waste management – the Indian perspective. J. Indian Assoc. Environ. Manag. **26**(2), 100–108 (1999)

Shekdar, A.V., Krshnawamy, K.N., Tikekar, V.G., Bhide, A.D.: Indian urban solid waste management systems – jaded systems in need of resource augmentation. J. Waste Manag. **12**(4), 379–387 (1992)

Siddiqui, T.Z., Siddiqui, F.Z., Khan, E.: Sustainable development through integrated municipal solid waste management (MSWM) approach – a case study of Aligarh District. In: Proceedings of National Conference of Advanced in Mechanical Engineering (AIME-2006), Jamia Millia Islamia, New Delhi, India, pp. 1168–1175 (2006)

Kumar, V., Pandit, R.K.: Problems of solid waste management in Indian cities. Int. J. Sci. Res. Publ. **3**(3), 1–9 (2013)

Applications of Recycled Sustainable Materials and By-Products in Soil Stabilization

Reem Alqaisi[1]([⊠]), Thang M. Le[1], and Hadi Khabbaz[2]

[1] School of Civil and Environmental Engineering,
University of Technology Sydney (UTS), Sydney, Australia
reemomar.alqaisi@student.uts.edu.au
[2] Geotechnical Engineering, School of Civil and Environmental Engineering,
University of Technology Sydney (UTS), Sydney, Australia

Abstract. The notion of sustainable and eco-friendly infrastructures is gaining impetus in the geotechnical engineering research field. Nowadays, employing recycled or waste materials, obtained from the natural sources, are widely used as an alternative method in construction in recognition of the green concept. It is evident that transport infrastructure projects require large amounts of materials and natural resources and consume large quantities of energy. On the other hand, large volumes of wastes are produced daily. In this paper, the beneficial effects of certain agricultural, domestic, industrial, construction, mineral and marine wastes in geotechnical applications, particularly in soil stabilization are discussed. Both methods of treatment are applied to improve the engineering properties of soil to make it suitable for construction. Another goal of this paper is to make a comparison between the effects of different types of waste materials for improving weak soil, and highlight the concept of sustainability to reduce energy consumption, carbon footprint, landfill cost and greenhouse gas emissions. Furthermore, many recycled materials and by-products are considered, and their advantages and drawbacks in soil stabilization are explained. In addition, the key results of many laboratory studies conducted on stabilized soil with waste materials are reported; and their effects on soil stabilization without or with a mixture of conventional stabilizers are discussed. The recent applications of wastes or recycled materials are also outlined. Finally, the future trend for employing recycled materials in infrastructure construction is presented.

Keywords: Soil stabilization · Sustainability · Waste materials · Industrial · Domestic and agricultural wastes

1 Introduction and Background

The concepts of waste materials are commonly described as material by-products rising from all human activities, which have no lasting value (Maghool et al. 2017). An enormous amount of raw materials, which exist in the world are consumed immensely in transport infrastructure facilitates such as roads, highways, bridges, airports, railways, waterways canals and terminals. As a result, the non-renewable natural materials are quickly exhausted (Bolden 2013; Correia et al. 2016). However, the reuse of

© Springer Nature Switzerland AG 2020
H. Ameen et al. (Eds.): GeoMEast 2019, SUCI, pp. 91–117, 2020.
https://doi.org/10.1007/978-3-030-34199-2_7

recycled materials in various types of infrastructure could profoundly reduce the need for natural materials, easing the pressure on landfills and stockpiles.

Figure 1 illustrates the types and amounts of waste materials from different sources, according to a report issued by Department of Sustainability, Water, Population and Communities in Australia during the year of 2010 and 2011. It is clear that the highest amount of these wastes (15 Mt) comes from Masonry materials, which include wastes as concrete, rubble and bricks. 70% of these wastes were recycled, and the rest were considered to be as a disposal. The second highest quantity was associated with the agricultural wastes and fly ash (14 Mt), for both of them, with 53% and 44%, respectively of recovery rate. It would be worth mentioning that 83% of the agricultural recovery rate was recycled and 17% was for energy recovery. The rest of the waste materials are shown in Fig. 1.

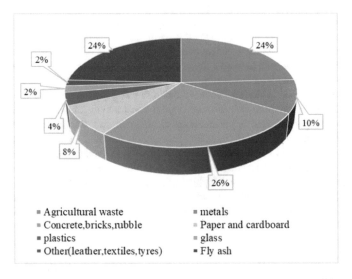

Fig. 1. Australia's total waste generation by material category management (millions of tonnes) in 2010–2011 (REC report 2011)

Most infrastructure activities leave behind significant environmental effects. For example, industries in the United States of America consumes annually over 40% of all raw stone, gravel and sand, 25% of all raw timber, 40% of energy, and 16% of the water (Chong et al. 2009). As a result, the concept of sustainability must be applied in accordance with the conservation of energy and original materials. Sustainability is a concept that originated from the direction of the world's attention towards better management of materials. Another definition of sustainability, given by the United Nations World Conference, held in 1987, is as follows: "management of resources such that current generations are able to meet their needs without affecting the ability of future generations to meet their needs". This definition has motivated many experts to increase interests in environmental resources and their management (Hill and Bowen 1997). In addition, one of the most important standards of sustainability in geotechnical applications is to reduce the use of concrete and soil by partial replacement of recycled waste materials and improving their characteristics.

Soil stabilization is highly recommended by engineers as a vital process to improve the performance of weak or marginal soil to achieve a higher level of stability. Chemical stabilization can be performed either by the use of recycled wastes and by-product materials as a stabilizer by themselves, or they may be used in combination with primary conventional stabilizing agents, such as lime or cement. Both methods of treatment are applied to improve the engineering properties of soil to make it suitable for construction. Soils are used in the most of civil engineering activities, the most important challenge is how to deal with problematic soils such as expansive soils. It is a term generally applied to any soil or rock material that has a potential for shrinking or swelling under changing moisture conditions. Severe damages could occur to the structures like light buildings, pavements, retaining walls, canal beds and linings. This leads to the necessity of determining the geotechnical properties of expansive soil in its natural state as well as when mixed with various proportions of waste materials, having beneficial effects on soil stabilization (Okagbue 2007; Rao 2007; Reddy and Muttharam 2001). Stabilization using waste materials is one of the techniques, employed to improve the engineering properties of expansive soils to make them suitable for construction (Sabat and Pati 2014). Lime, cement, pozzolanic fly ash, blast furnace slag, expanded shale and potassium chloride can be used for stabilizing expansive high plastic clayey soils to relatively shallow depths under footing and slabs for industrial foundations. One method of controlling the volume change is to stabilize the clay with an admixture to prevent the volume change (Amu et al. 2005). This paper proposes a detailed review of beneficial effects of certain industrial, agricultural, domestic, construction and marine wastes in geotechnical applications. It also discusses and analyses a large number of the experimental studies on waste materials for soil stabilization.

2 Comparison of Industrial Wastes for Soil Stabilization

Industrial solid wastes are usually generated by various industrial processes. They are usually placed in closed or open spaces to be disposed of around the perimeter of the plant. There are many industrial wastes that are used in soil stabilization. Table 1 illustrates 5 industrial wastes commonly used for soil stabilisation, including steel slage, fly ash, silica fume, red mud and lignin or pulp paper. It also shows the benefits and challenges of each sort of material referred to. As can be seen in this table, silica fume provides more advantages for stabilizing soil than any others while red mud appears as the least effective material. This might due to the fact that adding this fume in the soil makes their mixture more elastic through high bone strength, hence enhancing its durability. On the other hand, it is more difficult to obtain this state for red mud because of its weaker bonding from complex composition and impurity. Furthermore, red mud is prone to be hazardous to environment with their radioactive characteristics while silica fume is harmful to expiration because it has a great deal of majority of nano-sized particles. Other materials listed in Table 1 have more or less favourable features for soil stabilisation, namely, fly ash or pulp paper. Like silica fume, fly ash with its size and surface effects can decrease the volume of ash deposited in tailing dames. Pulp paper, nevertheless, is standing out with its sustainability when a great amount of paper waste can be used for engineering application. Fly ash and steel slag could be a part replacement for aggregates, but their danger to water can deter their stabilizing use in terms of environmental protection. Consequently, it recommended for silica fume due to their strength improvement in soil and fly ash for cheaper options while red mud can be used with its local availability. For details, each industrial waste in Table 1 is reviewed. After that, their comparison in the engineering properties of expansive soil stabilisation is shown. The involved experiments are with or without the presence of an activator (e.g. cement or lime) and include unconfined compressive strength (UCS), compaction properties, swelling properties and California bearing ratio (CBR), which are to assess the physical properties of weak soil after treatment and ensure the stabilised soil would be suitable to achieve the desired engineering requirements.

Table 1. Industrial waste materials and their specifications

Material	Advantages	Challenges	References
Steel slag	• Protection of natural aggregate	• High swelling • Water pollution	(Qasrawi et al. 2009; Sabat and Pradhan 2014; Yi et al. 2012)
Fly ash	• Decreased volume of ash deposited in tailing dams • Decreased carbon footprint	• Water pollution • Cost of transporting unit-value of coal fly ash • Lack of knowledge of potential ash uses and its Sensitivity to surface erosion	(Ahmaruzzaman 2010; Vaníček et al. 2016)
Silica fume	• High modulus of elasticity and Increased toughness • High bond strength and Enhanced durability • Superior resistance to chemical attack from chlorides, acids, nitrates and sulphates, etc. • High electrical resistivity and Low permeability	• Handling problem • Health hazards • High cost	(Khan and Siddique 2011; Mazloom et al. 2004; Sabat and Pati 2014; Siddique 2011)
Red mud	• Improved soil properties • Reduced foundation and building cost	• Difficulty in usage, due to its high alkalinity and water content	(Ashok and Sureshkumar 2014; Najar et al. 2012)

(*continued*)

Table 1. (*continued*)

Material	Advantages	Challenges	References
		• Complex composition and fine grained size • Limited application due to the presence of sodium and radioactivity • Environmental concerns	
Lignin/pulp paper	• Sustained natural resources • Decreased environmental pollution	• Health hazards • Unendurable use	(Balwaik and Raut 2011; Seyyedalipour et al. 2014)

2.1 Steel Slag

Steel slag considered one of the most common industrial wastes, which generated through either the steel-making process or by-product of the combustion process. This process produces different types of steel slag at different stages, such as electric arc furnace slag (EAFS) followed by ladle furnace slag (LFS). The high swelling potential is one of the main challenges appears to the slags applications (Hua-dong and Liu 2009; Maghool et al. 2017; Vaníček et al. 2016; Vaníček and Vaníček 2013). Many investigators conducted experimental programs to obtain the behaviour of stabilized soil using various types of steel slag (Goodarzi and Salimi 2015; Poh et al. 2006).

Activators such as lime and cement have a significant effect in improving the soil strength (Obuzor et al. 2011). Wild et al. (1998) delved into the positive impact of granulated blast furnace slag (GBFS) and its applications in soil durability enhancement, sulphate attack resistance, and chloride penetration resistance by partial replacement of lime in soil stabilization with GBFS. In a study conducted by Manso et al. (2013), the behaviour of clayey soil and its geotechnical properties by adding Ladle Furnace Slag (LFS). It was found that the behaviour of this combination provided the same impact on the soil and lime mixture by significantly increasing the compressive strength and reducing the plasticity index (IP) and free swelling. On the other hand, this research found that the durability index of the soil and LFS slag mixtures is higher than the durability index of the soil and lime mixtures. Another type of steel slag for soil stabilization, investigated by Akinwumi (2014), was an electric arc furnace (EAF) steel slag. The results clarified that the optimal value of (EAF) was 8%, and by adding this value the unsoaked CBR increased by 40% and UCS by 64%, while the liquid limit, plastic limit and plasticity index were reduced by 6.3%, 4.0% and 2.3%, respectively.

2.2   Fly Ash

An industrial by-product, fly ash is considered as a pollutant waste material, as it is generated from the combustion of coal in coal-fired electric power and steam generating plants. Due to the negative impact of fly ash on the environment, much research has been conducted to minimize the volume of ash and use it in soil stabilization (Sabat and Pati 2014). Various studies have been conducted to investigate the effects of adding fly ash and cement on different types of soil along with their geotechnical properties. Akinwumi (2014) found that the optimum percentage could approximately be 9% of cement and 3% of fly ash through which the geotechnical properties were highly improved. In studying the potential of stabilizing expansive subgrade soil, Zumrawi (2015) used various values of fly ash up to 20% (i.e. 0, 5 10, 15 and 20%), mixed with a constant amount of cement (5%). He found a significant improvement in the geotechnical properties of soil, such as its strength, durability, reduction in swelling in addition to its plasticity. Many researchers (Bhuvaneshwari et al. 2005; Bose 2012; Edil et al. 2006), however, found an improvement in the geotechnical properties after the utilization of fly ash even without adding cement into expansive soil. Wang et al. (2013) examined the strength and deformation behaviour of Dunkirk marine sediments with cement, lime, and fly ash. As expected, they observed that addition of activators increased the strength of stabilized sediments.

Many studies carried out to investigate the combination of fly ash and lime, another common conventional stabilizer. Sharma et al. (2012) used this combination for stabilization of clayey soil and found that the best result was obtained when the soil is stabilized by 20% fly ash and 8.5% lime, producing a sufficient strength for road subgrade. Mishra (2012) conducted another experiment on clayey soil by adopting fly ash as an additive to lime in soil stabilization; and he noticed that the optimum values for the additives lime and fly ash were 3% and 30%, respectively. Some other studies, conducted on expansive and black cotton soil, indicated improvement in the geotechnical properties of these soils when fly ash and other materials such as Magnesium and Aluminium chloride and polypropylene fibre were added (Radhakrishnan et al. 2014; Sabat and Pradhan 2014). On the stabilization of organic soils, Tastan et al. (2011) added fly ash as a stabilizer to study the degree of stabilization that will be obtained. When more additives like sand and tile waste are combined with fly ash for soil stabilisation, both soaked and un-soaked CBR of the admixture increase (Singh et al. 2014). In exploring the effect of Neyveli fly ash in stabilizing of shedi soil, Ramesh et al. (2011) found that the maximum strength of the stabilized soil could be obtained by adding 20% of Neyveli fly ash. Similarly, phosphogypsum can double the UCS of fly-ash blended with expansive soils (Krishnan et al. 2014).

2.3   Silica Fume

An industrial by-product of producing silicon metal or ferrosilicon alloys. One of its applications is soil stabilization and improve its geotechnical properties (Uzal et al. 2010). Different studies had been conducted to investigate the geotechnical properties of different types of soil when adding silica fume. The utilisation of silica fume not only reduces swelling pressure, permeability and enhance UCS but it also improves the durability of treated soil during the cycles of freezing and thawing (Kalkan 2009; Kalkan and Akbulut 2004). Moreover, a high strength and significant values of CBR obtained from the combination of lime and fly ash blended with a granular soil (Kalkan 2012). In connection to black cotton soil, (Negi et al. 2013) confirmed that the best percentage of silica fume should be 20% to obtain the 31 and 72% increase in UCS and CBR, respectively. With respect to lime combination, the lime-fume ratio of 5–9 to 10% in clayey soil stabilisation can improve the CBR, shearing capacity, and reduce the swell pressure and consolidation parameters (El-Aziz et al. 1981). On the stabilization of saline silty sand using lime and micro silica was experimented by Moayed et al. (2012). In connection with the difference between micro silica and lime or micro silica and cement on stabilization of low plasticity silty clay, It was noticed a significant increase in the strength when adding micro silica to lime, compared to the addition of cement (Bagherpour and Choobbasti 2003).

2.4 Red Mud

Red mud is a solid waste produced in the process of alumina production from bauxite following the Bayer process and it is considered hazardous due to its high pH (Ribeiro et al. 2011; Samal et al. 2013). Several studies had been made on the effects of using red mud on soil stabilization. The UCS value increased by 2 times compared with the untreated soil when the combination of red mud and sodium silicate blended with black cotton soil (Mane and Rajashekhar 2017). Furthermore, the CBR increased by 2% at the optimum dosage of 8% Sodium silicate. In case of blending with conventional stabilizer, a mixture of cement – red mud (CRM) resulted in a turn in the soil groups from high-plasticity soil group to low-plasticity soil group (Kalkan 2006). Moreover, the possibility of replacing the soil by red mud on treating with lime was investigated (Gayathri 2016).

2.5 Pulp Paper

After being lightened, pulping wood is spread out into pieces in order to make paper. Then, it is processed through several chemical stages to give it certain properties such as turning it into white colour through the bleaching chemicals stage, enabling it to be in various colours (Zavatta 1993). In order to discover the effects of paper on the properties of various soils, tests on several soil mixtures were carried out. The expansive soil was treated with paper mill ash to test its UCS, CBR, Atterberg limits, MDD and OMC, all of which were improved (Byiringiro 2014). With the optimum content of paper mill ash at 20%, the CBR growth is observed obviously from 2.8 to 64.4%, indicating the highest positive effect of pulp paper on soil strength reinforcement. Moreover, paper mill ash from a multi-fuel boiler was a good choice to use in expansive soil treatment. Eroglu et al. (2004) found out that lime mud from the paper industry enhances the geotechnical properties of soil used to stabilize forest road superstructures.

2.6   Engineering Properties of Industrial Recycled Waste Materials Treated Expansive Soils

The improvements achieved of the unconfined compressive strength when the untreated soil stabilized by using some of the industrial wastes without any conventional stabilizer (lime/cement) or with combinations of lime/cement treated expansive soil is shown in Fig. 2. The figure depicts that the addition of industrial solid wastes has resulted in improving the UCS as investigated by various researchers. At the optimum state, there is at least 29% increase in the UCS test result in case of 30% silica fume. In addition, at the best state, it is clear that the addition of 20% Neyveli ash has the maximum improvement on UCS, as it is around 13.75 times more than the UCS before improvement. As for the mixtures of 8% steel slag, 30% silica fume, 20% silica fume,

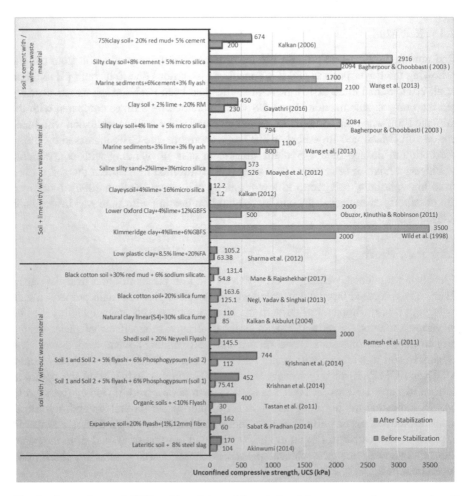

Fig. 2. Comparison of UCS achieved by adding industrial wastes to the untreated soil with/without lime or cement treated soil

and 30% red mud) to soils, these materials have a relatively equal effect on soil improvement as shown in the UCS results. The improvement obtained due to the addition of industrial waste materials at the optimal dosage to lime/cement stabilization also shown in the Fig. 2. It can be seen that the addition of industrial waste produced better results than only using conventional stabilizer, such as lime or cement. Stabilization using granulated blast furnace slag (GBFS) produce excellent strength as well as swell control and it was adopted with highly plastic, expansive clays. This figure has been shown positive results by using combinations of lime/cement with industrial wastes at its optimal dosage.

The improvement of soaked California bearing ratio (CBR) obtained when the untreated soil stabilized by using some industrial wastes material without any conventional stabilizer or with a combination of lime treated expansive soil is shown in Fig. 3. In case of using 8% steel slag on stabilizing laterite soil, any increase in the steel slag content contributes in lowering CBR values as they are strongly correlated on soil because the value of the soaked CBR is decreased. This figure indicates that the combinations of 20% fly ash with 1% fibre in the length of 12 mm, 63:27:10 fly ash with 9% tile waste, 20% silica fume, and 30% red mud with 6% sodium silicate have relatively equal effects on increasing the CBR value 4%, 3%, 1% and 2%, respectively. In fact, paper mill ash has the most effect on growing the CBR from 2.8% to 64.4%. Consequently, there are instances of CBR rising in the case of paper mill ash and just by 1% in the case of silica fume. This indicates that the development of strength is

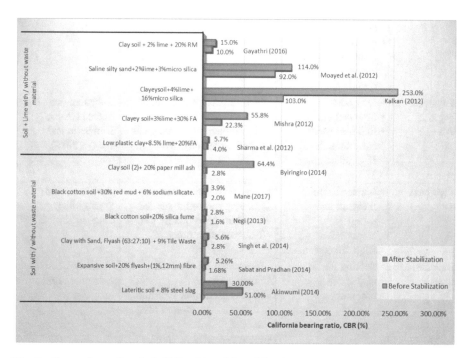

Fig. 3. Comparison of soaked CBR achieved by adding industrial wastes to the untreated soil with/without lime treated soil

influenced by the soil type and the nature of the industrial waste. Moreover, as shown in Fig. 3, when the industrial wastes added to the lime-treated expansive soil, a noticeable increase in the CBR values were obtained.

3 Comparison of Agricultural Wastes for Soil Stabilization

Agricultural wastes are obtained from the agricultural resource. The processing of various agricultural products including sugarcane bagasse ash, rice husk ash, bagasse fibre, groundnut shells, whey, wood ash, burnt olive waste and coconut coir fibre generates them. Table 2 illustrates 7 agricultural wastes that have been used in soil stabilization and it also shows the benefits and challenges for each sort of stabiliser. A number of investigations, conducted on the stabilization of soil using agricultural wastes, are described below.

Table 2. Agricultural waste materials and their specifications

Materials	Advantages	Challenges	References
Coconut coir fibre	• Low cost • Easy to use or handle • Resistant to fungi and rot • Excellent insulation against temperature and sound	• Changed chemical composition of fibres because of immersion in water, saturated lime, and sodium hydroxide leading to loss of their strength	(Ali et al. 2012; Nadgouda 2014)
Sugarcane bagasse ash	• Low cost • Low co2 emission • Reduced energy consumption. • High silica content 87%	• Low impact on strength of soil • The problem of stocking raw material for an extended time	(Schettino and Holanda 2015; Srinivasan and Sathiya 2010)
Rice husk ash	• Reduced carbon footprint and energy consumption • High silica content	• A suitable incinerator and grinding method is needed to obtain good quality ash • Transportation problem	(Al-Khalaf and Yousif 1984; Ganesan et al. 2008)
Bagasse fibre	• Solving its disposable problem	• Incompatibility between fibres • Poor resistance to moisture • High concentration of fibre defects • Fibre degradation during processing	(Aggarwal 1995; Aziz et al. 1981)

(*continued*)

Table 2. (*continued*)

Materials	Advantages	Challenges	References
Peanut shells	• Low capital cost per tonne production compared to cement • Reduced pollution and CO2 emission • Conserved limestone deposits • Good Pozzolanic material, which reacts with calcium hydroxide	• Little potential use • Not meeting the criteria for soil properties	(Bell 1996; Oriola and Moses 2010)
Wood ash	• Minimized energy consumption and carbon footprint • Low cost • Improved soil alkalinity as fertilizer • Being able to capture several waterborne contaminants	• Being prone to cracking, deforming or insects due to its low corrosion resistance • Poor stability	(Bohrn and Stampfer 2014; Cheah and Ramli 2011)
Burnt olive waste	• Saving natural material • Low cost Reduced environmental pollution	Inverse impact on soil properties if exceeding the optimum percentage	(Al-Akhras and Abdulwahid 2010; Nalbantoglu and Tawfiq 2006)

As can be seen in this table, sugarcane bagasse ash, bagasse fibre, and rice husk ash have distinctive properties due to their high percentage of silica. The percentage of silica in bagasse ash is approximately 87%, while in rice husk ash is about 90–98%. This great amount of silica can enhance the engineering characteristic of expansive soils by improving the efficiency of pozzolanic reaction. Meanwhile, peanut shell ash contains a small percentage of silica, which could limit its potential for meeting the criteria of the soil strength. Utilization of these agricultural wastes in soil stabilization can reduce the carbon footprint and energy consumption as well as solve its disposable problem. Particularly, coconut fibre proves to be a good reinforcement material due to it's resistant to fungi and rot and its excellent insulation against temperature and sound. Nevertheless, there is an inverse impact if exceeding the optimum percentage of coconut shell. The same impact on soil properties is noticed with adding a high percentage of olive wastes. Furthermore, for improving soil alkalinity, it was suggested to use wood ash as a stabilizer (Brown et al. 1998). In addition, it is recommended to employ rice husk ash in soil stabilisation due to its great amount of silica and agricultural fibres for soil strength reinforcement. However, despite the advantages of bagasse fibre in soil strength reinforcement, the durability of this mixture should be improved with fibre treatment (e.g. soaking fibres in sodium hydroxide before soil mixing).

3.1 Sugarcane Bagasse Ash

Sugarcane bagasse is a fibrous waste product of the sugar refining industry. When the juice is extracted from the cane sugar, the solid waste material is known as bagasse. When this waste is burned, it gives ash called bagasse ash. Bagasse ash mainly contains aluminium ion and silica. It is a fibrous material with the presence of silica ($SiO2$) can be used to improve the existing properties of black cotton soil (Sabat and Pati 2014; Srinivasan and Sathiya 2010). Several studies were performed on the effects of Bagasse ash (BA) mixed with lime and cement activators on the geotechnical properties of different types of soils. Many experiments were conducted on mixing BA with lime (Manikandan and Moganraj 2014; Osinubi et al. 2009; Surjandari et al. 2017). On the stabilization of compacted soil blocks, Alavéz-Ramírez et al. (2012) concluded that 10% lime with 10% BA enhanced the mechanical properties of compacted soil blocks. In connection with lateritic soil, a comprehensive experimental programme conducted to show the effect of blending lime and BA with lateritic soil (Ochepo et al. 2015; Sadeeq et al. 2015). In order to investigate the effectiveness of lime and bagasse ash admixture in reinforcing the sub-grade of a flexible pavement on the expansive soil, Sabat (2012) concluded that 8% BA and 16% lime sludge were the optimum values at which the best stabilization could be obtained. Mu'Azu (2007) added up to 8% bagasse ash with up to 4% cement by weight of the lateritic soil to investigate the influence of British Standard Light (BSL), West African Standard (WAS) and British Standard Heavy (BSH) compactive effort. Three different BA contents of 2%, 4%, and 8% were blended with two different cement contents by Lima et al. (2012). For black cotton soil stabilisation, different amounts of BA were added to the soil, and as a result, the optimum value was found to be 6% replacement of bagasse Ash. The best stabilization was obtained at this value, as MDD increased by 5.8%, CBR increased by 41.52% and UCS increased by 43.58% (Kharade et al. 2014).

3.2 Rice Husk Ash

After rice husk is obtained from paddy, it is disposed of by various methods, either "by dumping it in an open heap near the mill site or on the roadside to be burnt later". Being rich of siliceous material, means that it has pozzolanic properties (Rao and Chittaranjan 2011). Many researchers studied the effect of using rice husk ash (RHA) with cement in stabilizing different types of soil (Roy 2014). Alhassan and Mustapha (2007) added this mixture on lateritic soil. On the same type of soil, Rahman (1987) used the same mixture of cement and RHA in order to find out its potential to be used in highway construction. He figured out that for the sub-base materials, the best result was 6% RHA and 3% cement. For base-materials, he concluded that the best result was 6% for both cement and RHA. For investigating the stabilization of residual soils, Basha et al. (2005) had used cement and RHA. The combination of cement and RHA would achieve a given strength with only a little amount of cement as compared to adding cement alone. In stabilization of high plasticity clay, Roy (2014) delved into the effect of RHA and cement mixture. Other studies were carried out using RHA, lime and calcium chloride and polypropylene on certain types of soil. On clayey soil, Muntohar et al. (2000) added this mixture to find out its geotechnical properties. Sabat (2012) analysed using polypropylene fibre in addition to the RHA and lime mixture to stabilize the expansive soil. Sharma et al. (2008) implemented an experiment on the effects of adding calcium chloride, lime and RHA.

3.3   Bagasse Fibre

As a reinforcing component for expansive soil stabilization, bagasse fibre is produced from the sugar cane industry. After the sugar cane is crushed for juice extraction, bagasse fibre is obtained (Onésippe et al. 2010). A number of tests were conducted to uncover the effects of bagasse fibre on the geotechnical properties of stabilized expansive soils. A study by Oderah (2015) on sandy and clayey soils was conducted by adding various bagasse fibres content. He found out that there is a relation between bagasse fibre content and the shear strength parameters values of the soils, showing that the more the bagasse content is, the more the shear strength parameters values are, and the optimum content of bagasse fibre was 1.4%. The combination of hydrated lime and bagasse fibre proved to be more efficient to enhance soil properties than adding bagasse fibre by itself. This is what was emphasized by Dang et al. (2016) who had found that the optimum values are 0.5% bagasse fibre with 7% lime at which the expansive soil properties were improved. With the increase of curing time, bagasse fibre and hydrated lime, there was an observable decrease in the linear shrinkage of treated expansive soil.

3.4   Groundnut Shell Ash

Many researchers studied the feasibility of using groundnut shell ash (GSA) to stabilize black cotton soil. The milling of groundnut results in groundnut shell, which is considered as a waste. Because the accumulation of the groundnut shells has a negative impact on the environment and because of their low cost, they can be turned into ash in order to be utilized in stabilizing poor soil. Several tests were conducted (Ijimdiya et al. 2012; Otoko and precious 2014) in order to check the geotechnical properties for black cotton soil. It was concluded that UCS depends on the curing period of the soil. Peak UCS at 7 days failed to meet the acceptable requirement for the base, sub-base and subgrade, while with a longer curing time, gradual strength development was noticed. From this observation, for better stabilization with groundnut shell ash, it is recommended to mix it with a stronger stabilizer.

3.5   Wood Ash

Whether burned in a home fireplace or an industrial power plant, the combustion of wood produces organic and inorganic powder known as wood ash. Being rich in potash make it useful for gardeners. Several tests have been made on different types of soils to investigate the effectiveness of wood ash in increasing soil strength. The stabilization of clay soil by using wood ash was evaluated by Okagbue (2007). Adding different proportions of wood ash, 1–25% weight percent on four-clay soil specimens, Barazesh et al. (2012) conducted an experiment to check the soils' Atterberg limits. While Barazesh et al. (2012) conducted their experiments on a clay soil to check the Atterberg limits; Šķēls et al. (2016) implemented their experiment on the natural sand (SA) soil to check its load-bearing capacity. The test was also made to check wood fly ash potential as a good hydraulic binder with sand.

3.6   Burnt Olive Waste

After pressing olives and extracting them to produce oil, the by-product olive waste is left behind. The olive cake residue is continuously increasing in the Mediterranean countries, and it is predicted to double in an amount within 10–15 years. This quantity of the by-product olive cake could be beneficially used as a soil stabilizer (Nalbantoglu and Tawfiq 2006; Rao and Chittaranjan 2011). Many researchers explored the effectiveness of using burnt olive waste in stabilizing expansive soils as a way to solve the problem associated with the increase of olive waste. Attom and Al-Sharif (1998) conducted a chemical analysis to recognize the elements constituting the olive waste after burning at 550 °C as well as an array of geotechnical tests were conducted. It was concluded that there was a potential to use this waste as a soil stabilizer. Nalbantoglu and Tawfiq (2006) found that adding 3% burned olive waste into the soil causes a reduction in the plasticity and an increase in the UCS. Based on these results, the use of olive waste in soil stabilization is greatly beneficial for the environment as it achieves the concept of sustainability. After the olive waste was burned at 5508C, different proportions of this burned olive 1–10% were added by Mutman (2013) to the bentonite clay.

3.7   Coconut Coir Fibre

Coconut coir belongs to the hard-structural fibres group, and it is taken from coconut husk. It is an organic fibrous material containing 40% lignin and 54% cellulose. Because of its high lignin content, coir has a potential to be used as reinforcing the material in the soil (Khatri et al. 2016). Many researchers have shown that coir fibre reinforcement can significantly improve the engineering properties of soil. Abhijith (2015) conducted an experimental program to study the effect of coir textiles on CBR strength of soil subgrade and to determine the ideal position to place the coir geotextiles. The effectiveness of adding coal ash and coconut coir fibre in local soil (silty sand) was explored by Singh and Arif (2014). The optimum percentages of soil, coal ash and coir fibre mix were reached at 79.75:20:0.25 by weight of dry soil. At these values, all properties of the soil were greatly improved. Anggraini et al. (2015) investigated the effect of coir fibre content and curing time on the tensile and compressive strength of soft soil treated by lime. The test results indicated that compressive strength less sensitive to lime and coir fibre stabilization compared to tensile strength. Another conventional common stabilizer (cement) adopted by Yadav and Tiwari (2016) to study the combined effect of coconut fibre and cement in increasing the strength of clay-pond ash (PA) soil.

3.8 Engineering Properties of Agricultural Recycled Waste Materials Treated Expansive Soils

The improvements achieved of the unconfined compressive strength when the untreated soil stabilized by using some of the agricultural wastes without any conventional stabilizer (lime/cement) or with combinations of lime/cement treated

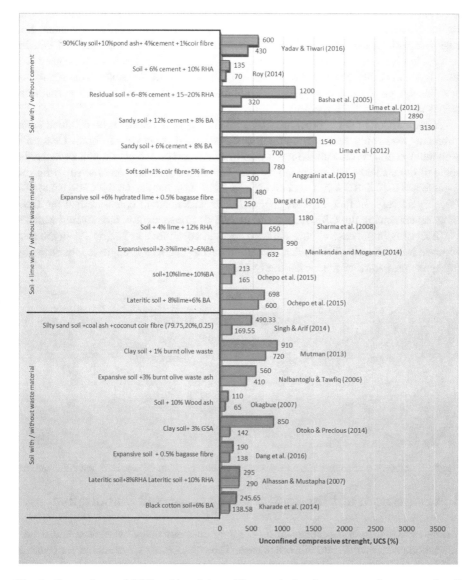

Fig. 4. Comparison of UCS achieved by adding agricultural wastes to the untreated soil with/without lime or cement treated soil

expansive soil is shown in Fig. 4. The figure depicts that the addition of agricultural solid wastes has resulted in improving the UCS as investigated by various researchers. At the optimum state, there is at least 2% increase in the UCS test result in the case of 8% RHA. In addition, at the best state, it is clear that the addition of 20% coal ash with 0.25% coconut coir fibre to the silty sand soil and 3% groundnut shell ash to the clay soil have the maximum improvement on UCS, as it is around 3 times more than the UCS before improvement. As for (6% bagasse ash, 0.5 bagasse fibre, 3% burnt olive waste ash, and 1% burnt olive waste), these materials have a relatively equal effect on soil improvement as shown in UCS results. The improvement obtained due to the addition of agricultural waste materials at the optimal dosage to lime/cement stabilization also shown in Fig. 4. This figure depicts that the addition of agricultural waste produced better results than the conventional stabilizer only. Stabilization using rice husk ash and bagasse ash produce excellent strength with both of lime and cement stabilization as well as swell control. Consequently, RHA and BA were adopted frequently as an additive with lime and cement, based on proven results. This figure has been shown positive results by using combinations of lime/cement with agricultural wastes at its optimal dosage.

A comparison in the improvement of soaked California bearing ratio obtained when stabilizing soil using some agricultural wastes material is shown in Fig. 5. Using 5% groundnut shell ash on stabilizing clay soil has the lowest effects on soil because the value of the soaked CBR increased by 1%. This figure indicates that the soil with 6% bagasse ash to10% RHA has a relatively equal effect on increasing the CBR value 9%, 7%, respectively. In fact, 20% of coal ash with 0.25 coconut coir fibre had the most effect on enhancing the CBR from 5.6% to 19.5%. Consequently, there were instances of CBR rising in the case of coal ash with coconut coir fibre and just by 1% in the case of groundnut shell ash. This indicates that the development of strength is influenced by soil type and nature of the agricultural waste.

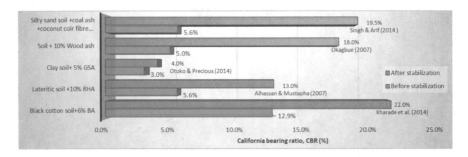

Fig. 5. Comparison of soaked CBR achieved by adding agricultural wastes to the untreated soil

4   Comparison of Miscellaneous Wastes for Soil Stabilization

Miscellaneous wastes coming from different sources such as construction and demolition, domestic, marine and mineral wastes. Table 3 illustrates 7 miscellaneous wastes that have been used in soil stabilization and it also shows the benefits and challenges

Table 3. Miscellaneous waste materials and their specifications

Materials	Advantages	Challenges	References
Construction waste			
Recycled asphalt pavement	• Reduced landfill cost • Reduced carbon footprint • Reduced demand for natural aggregate	• Lack of laboratory and field performance data • Lack of proper standard • Lack of located recycling facilities	(De Bock and Gonçalves 2005; Gomes Correia et al. 2016; Taha et al. 2002)
Domestic waste			
Eggshell powder	• Conserving natural lime • Low cost • Reduced carbon footprint • Increased soil density, cohesion and frictional resistance	• Adverse impact on soil properties if exceeding the optimum percentage	(Erfen and Yunus 2015; Hut 2014; Pliya and Cree 2015; Raji and Samuel 2015; Rao and Chittaranjan 2011; Yerramala 2014)
Tyre rubber	• Reduced carbon footprint and energy • Low cost of raw material	• Negative impact on soil stabilization in case of tires abrasion	(Cairns and Kenny 2004; Huang et al. 2006)
Glass cullet	• Improved soil properties • Reduced disposable wastes.	• Adverse impact on soil properties if exceeding the optimum percentage	(Boraste and Ikara 2015; Olufowobi 2014)
Carpet fibre	• Reduced needs for landfilling • Reduced energy carbon footprint •Increased in the internal cohesion of soil	• Only limited types of fibres could be used in soil stabilization	(Mirzababaei et al. 2009; Wang 1999; Wang et al. 1994)
Marine waste			
Sea shells	• Conserving virgin materials • Reduced pollution and energy • Enhanced soil properties Being rich source of several bioactive compounds and materials, such as calcium, chitin, pigments, and proteins	Adverse impact on soil properties if exceeding the optimum percentage	(Hanif et al. 2014; Othman et al. 2013)

Mineral waste

Quarry dust	• Low cost by Reducing requirement of landfill area • Solving the problem of natural sand scarcity • Improved soil properties Variation in water content does not seriously affect its desirable properties	Difficulty in cutting grove in casagrandes apparatus when applying the liquid limit test	(Balamurugan and Perumal 2013; Soosan et al. 2005)

for each sort of the mentioned material. Some of the research conducted on the stabilization of soil using miscellaneous wastes are described below. As can be seen in this table, eggshell powder and seashell powder have a high percentage of CaO. Consequently, these materials can be used for increasing the efficiency of pozzolanic reaction as well as conserving the natural lime. An adverse impact was noticed by adding high dosage from these materials. On the other hand, due to the granular properties for both of recycled asphalt pavement and the shredded tyre rubber, it could be used as aggregate as well for transportation construction such as road subgrade or subbase and embankment fill. However, tyre rubber has a negative impact on soil stabilization in case of tires abrasion whereas, a noticeable improvement in the properties of the stabilized soil is by addition of quarry dust and glass cullet. Furthermore, the low cost of the process of obtaining fibres from carpet waste makes it a good choice for soil stabilization and increasing the internal cohesion of soil. However, limited types of fibres could be used in soil stabilization.

4.1 Construction Waste Materials (Recycled Asphalt Pavement)

Building and construction industry generates waste known as the construction and demolition waste (C&D). Examples of this waste include bricks, tiles, masonry, cement, timber, metals, plastics and cardboard. Much research has been conducted on the C&D potential to be used in soil stabilization. After recycling waste obtained from the end-of-life asphalt and concrete pavements into aggregate or pulverized, it is stabilized into full or partial depth reclamation bases by adding cement or other additives as hydraulic binders (Puppala et al. 2011). Maintenance works on asphalt work pavements results in a significant amount of reclaimed asphalt which is caused by "milling of asphalt road layers, by crushing of slabs ripped up from asphalt pavements, lumps from slabs, and asphalt from reject and surplus production" (De Bock and Gonçalves 2005). The suitability of recycled asphalt pavement (RAP) stabilized with cementitious sawdust ash as pavement material was studied by Osinubi et al. (2011). Moreover, Taha et al. (2002) also conducted an extensive laboratory evaluation of cement stabilized RAP and RAP-virgin aggregate blends as base materials.

4.2 Domestic Waste Materials

From post-Consumer commercial and household waste, domestic solid wastes are generated. Eggshells powder, tyre rubber, glass cullet and waste carpet fibres are examples of domestic wastes.

Eggshell is a waste material generated from domestic sources such as poultries, hatcheries, homes and fast food centres. Large amounts of Eggshells pose a significant environmental concern if not recycled or discarded properly (Pliya and Cree 2015). Due to its similar chemical composition to that of lime, it can be used as a partial replacement for industrial lime if subjected to adequate scrutiny (Rao and Chittaranjan 2011). As a result, they could be efficient for soil stabilization. Different studies had been implemented to discover the effects of adding eggshell powder (ESP) with different types of soils. Amu et al. (2005) studied the effects of adding lime with ESP on the stabilizing potential of the soil while Barazesh et al. (2012) added ESP by itself. Nyankson et al. (2013) implemented their experiment on two different expansive soil samples from Dodowa (DD) and Adalekope (AD) in Ghana. The optimum percentages of ESP were observed to be at 4% and 8% at which the swell-shrinkage behaviour improved. Soundara and Vilasini (2015) probed the effect of ESP on the properties of clay. On the lateritic soil, Olarewaju et al. (2011) had studied the ability of eggshell and cement as subgrade materials for road construction. It is concluded that eggshell can be used as a stabilizer for subgrade materials. Paul et al. (2014) used a combination of quarry dust (QD) and eggshell powder to stabilize clayey soil. Muthu Kumar and Tamilarasan (2014) investigated the use of eggshell powder to stabilize the soil. It was noticed that for every addition of 0.5% eggshell powder to the soil, the UCS increased approximately by 25%. Another additive, which called common salt used to study its effect on the properties of eggshell stabilized lateritic soils by Amu and Salami (2010). The combustion of ESP in a muffle furnace at elevated temperatures produces eggshell ash (ESA). Few experiments have been conducted to detect the possibility of using eggshell ash as an auxiliary addendum in the stabilization of expansive soils (James and Pandian 2016; James et al. 2017; Okonkwo et al. 2012).

Significant environmental concerns are associated with waste tires if not recycled properly. In civil engineering, there are many applications for the recycled waste tyres and rubber crumbs such as asphalt paving mixtures, playground surfaces and concrete for rigid pavements. Small particles of waste tyre rubber were added to the expansive soil to figure out its swelling potential. It was noticed that adding this waste into the expansive soil significantly improved soil behaviour. Because of the lower specific gravity of the crumb rubber, at its optimum value, the maximum dry density of mixes decreases. Hence, it was concluded that expansive soil-rubber mixture (ESR) is more compressible than the untreated soil. Hoan and Ming (2010) studied the mechanical and chemical properties of the cement-rubber chip to stabilize soft clay soil (kaolin) by adding shredded rubber as the reinforcement material with a binding agent, cement. They concluded that the efficiency of adding rubber and cement into the soil is much better than adding rubber by itself.

An amorphous non-crystalline material, the glass is usually brittle and optically transparent. In addition of using glass as a raw material to be recycled into other glass containers, it may be used in some engineering applications. As being produced through supercooling processes, it is composed mainly of silicon dioxide (sand) and sodium carbonate. Researchers found out that the crushed glass has many positive physical properties, high permeability, high crushing resistance, small strain stiffness. These properties enable the crushed glass to be used in soil stabilization (Rao and Chittaranjan 2011). In the investigation into the combined effect of glass waste and cement for expansive soil stabilisation, the optimum glass-cement of 20 to 8% gave the highest UCS and CBR of 1.2 MPa and 53.8% respectively after 7 days for curing (Ikara et al. 2015). For black cotton soil, 5% content of glass powder produced the maximum CBR value while 10% glass proportion enhanced the shearing parameters highest (Olufowobi et al. 2014), which indicates that glass powder can be the suitable binder for clay stabilisation. Concerning the laterite soil, additional fly ash is added in the glass-treated soil and an optimum percentage of ash-glass combination can be found out. Based on the study of (Boraste and Sharma 2014), the ratio of 7% glass power to 20% fly ash can improve the soaked and un-soaked CBRs at their highest values. Consequently, glass waste can be effective soil stabiliser because of its outstanding improvement in the engineering behaviour of expansive soil. Carpet fibres are recovered by shredding carpets, and it acts as tensile reinforcing elements in the soil when blended with conventional equipment. The shredded carpet waste fibres are proposed to be up to 70 mm long as shown by field trials (Wang 2006). Many researchers had investigated the effects of using carpet fibre to increase the strength of different types of soil. In the form of short strips of carpets, Ghiassian et al. (2004) had examined the effects of using recycled carpet waste on a fine sandy soil. For reinforcing soil, another experiment, using carpet waste fibre by Wang (2006) was conducted. Mirzababaei (2009) using four different surplus carpet fibre types for studying some geotechnical properties of cohesive soil.

4.3   Marine Waste Material (Seashells)

From the marine environment, marine wastes are generated, including seashells which are a hard, protective outer layer created by an animal that lives in the sea. There is no sufficient research that could cover the use of seashells in soil stabilization. Seashells is a hard exoskeleton of molluscs naturally available on the seashores, Seashells contain about 90% of calcium carbonate, which is a major component in lime. It was studied the effects of seashells as a stabilizing agent in the black cotton soil. Based on the results, it was noticed that the UCS and CBR had increased by increasing the seashells content (Mounika et al. 2014).

4.4   Mineral Waste Material (Quarry Dust)

Mineral solid wastes are the waste generated from the extraction of ores and minerals. Some of the research conducted on stabilization soil by using Quarry dust. After crushing rubble, the quarry dust is obtained as an aggregate waste. This dust material could treat soil subgrade/subbase without additional additives like lime or cement, but it still produces reasonable improvements. At the 40% in quarry dust content, the CBR of treated samples was read at the highest value, indicating this percentage is the optimum value of material content for soil reinforcement. The dust addition also helps reduce the liquid limit, thus decrease the plasticity index. Repeatedly, the optimal content of quarry dust at 40% was obtained for the optimum CBR value when the material was mixed with black cotton soil (Chansoria et al. 2016; Kumar and Biradar 2014). However, this percentage should be 45% to result in the highest UCS and CBR as well as the shearing characteristics of stabilised soils (Sabat and Bose 2013). With reference to the combination of quarry dust and lime for their effects on engineering properties of expansive soil, the optimum ratio of 40% dust is combined with the increasing content of lime from 2 to 7% make the quarry dust-treated soil more durable than admixtures without lime (Sabat 2012).

4.5 Engineering Properties of Miscellaneous Recycled Waste Materials Treated Expansive Soils

A comparison in the improvement of the unconfined compressive strength obtained when the untreated soil stabilized by using some of the miscellaneous wastes without any conventional stabilizer or with combinations of lime/cement treated expansive soil is shown in Fig. 6. The figure depicts that the addition of miscellaneous solid wastes has resulted in improving the UCS as investigated by various researchers. At the optimum state, there is at least 3 times increase in the UCS test result in mixing 3% ESP in the expansive soil, 6% ESP in the clay soil, and quarry dust (1:2) with 45% Fly

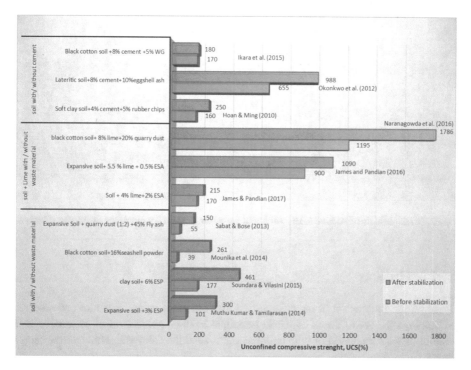

Fig. 6. Comparison of UCS achieved by adding miscellaneous wastes to the untreated soil with/without lime or cement treated soil

ash. In addition, at the best state, it is clear that the addition of 16% seashell powder has the maximum improvement on UCS, as it is around 7 times more than the UCS before improvement of soil. The improvement obtained due to the addition of miscellaneous waste materials at the optimal dosage to lime/cement stabilization also shown in Fig. 6. This figure depicts that the addition of miscellaneous wastes produced better results than the conventional stabilizer only. Furthermore, positive results were obtained by using combinations of lime/cement with miscellaneous wastes at its optimal dosage.

A comparison in the improvement of soaked California bearing ratio is obtained when the untreated soil stabilized by using some miscellaneous wastes material is shown in Fig. 7. Using 40% quarry dust on stabilizing soil has the lowest effects on soil because the value of the soaked CBR was increased by 2%. This figure indicates that the compounds of 90% RAP with 10% sawdust ash, 20% seashell powder, quarry dust (1:2) with 45% Fly ash, and 40% quarry dust in the black cotton soil have a relatively equal effect on increasing the CBR value 3%, 7%, 4% and 5%, respectively. In fact, 90% RAP with 10% sawdust ash has the most effect on growing the CBR from 23% to 26%. Consequently, there are instances of CBR rising in the case of RAP with sawdust ash and just by 2% in the case of quarry dust.

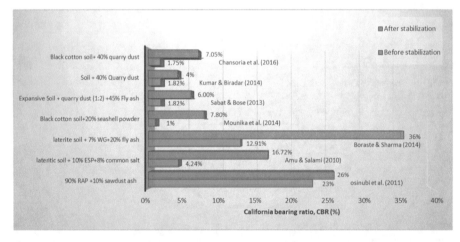

Fig. 7. Comparison of soaked CBR achieved by adding miscellaneous wastes to the untreated soil

5   Conclusions

Based on the integration of many investigations available in the literature and research on employing several types of waste materials in the stabilization of expansive soil, without or with inclusion of conventional stabilizers, the following conclusions can be drawn:

- The stabilization of expansive soil using different types of wastes, whether by itself or with activators (lime or cement), improves the geotechnical properties of soil. In addition to the methodology of investigation, the performance of wastes depends on certain factors including the type of soil, the nature and amount of waste as well as the type and amount of the conventional stabilizers.

- This study has attempted to provide a wide range of stabilizing effects using industrial, agricultural and miscellaneous wastes with weak soils. Industrial wastes consist of steel slag, fly ash, silica fume, red mud and pulp paper. Meanwhile, agricultural wastes include bagasse ash, rice husk ash, bagasse fiber, groundnut shell ash, wood ash, burnt olive waste and coconut coir fiber. Moreover, miscellaneous wastes comprise of construction wastes and different types of domestic wastes.

- Literature indicates that waste materials can increase the stability of soil similar to conventional stabilizers such as lime/cement. However, combination of these two in soil stabilization generates even better results in many conditions. As a result, the amount of conventional stabilizers in treatment of problematic soils will gradually be reduced. Accordingly, the carbon footprint of manufacturing of such materials will be reduced.

- Among industrial wastes, silica fume is described as a promising material for soil reinforcement. The UCS values increased 3 times compared with the UCS of lime treated soil. Fly ash is also considered as an effective reactive agent for soil stabilization particularly when combined with lime. For class F fly ash an activator (either cement or lime) is essential. Referring to CBR results, micro silica has been the most effective by product, added to lime treated soil. The same behaviour also noticed when fly ash and silica fume blended with the cement stabilized the soil. Based on the literature, Nivel fly ash can provide the highest strength, as high as 2000 kPa. Others by products such as steel slag, red mud and pulp paper produced less enhancement in soil strength.

- Regarding agricultural wastes, with introduction of lime or cement, a significant improvement on soil strength and CBR were observed, when bagasse ash and rice husk ash mixed with the soil. Furthermore, coconut coir fibre and bagasse fiber have shown their effectiveness in soil reinforcement. Materials such as olive waste and groundnut shell ash have exhibited less enhancement effect on soil strength.

- Regarding miscellaneous wastes, it can be noted that limited studies were conducted on the effect of applying these wastes combined with of lime or cement on treated soil properties. However, eggshell powder and seashell powder are described as promising materials for enhancement the soil strength or replacement of lime due to their high content of calcium oxide. Referring to UCS test results, quarry dust yielded better improvement compared to recycled asphalt pavement or glass cullet.

- By and large, studies on using wastes for soil stabilization offer an insight into their positive impacts on expansive soil treatment.

- Finally, the application of wastes and by products in soil stabilization can significantly contribute in reducing the costs of pavement construction, which would be of economic importance in both underdeveloped and developing countries.

References

Aggarwal, L.K.: Bagasse-reinforced cement composites. Cement Concr. Compos. **17**(2), 107–112 (1995)

Ahmaruzzaman, M.: A review on the utilization of fly ash. Prog. Energy Combust. Sci. **36**(3), 327–363 (2010)

Al-Akhras, N.M., Abdulwahid, M.: Utilisation of olive waste ash in mortar mixes. Struct. Concr. **11**(4), 221–228 (2010)

Al-Khalaf, M.N., Yousif, H.A.: Use of rice husk ash in concrete. Int. J. Cem. Compos. Lightweight Concrete **6**(4), 241–248 (1984)

Ali, M., Liu, A., Sou, H., Chouw, N.: Mechanical and dynamic properties of coconut fibre reinforced concrete. Constr. Build. Mater. **30**, 814–825 (2012)

Amu, O., Fajobi, A., Afekhuai, S.: Stabilizing potential of cement and fly ash mixture on expansive clay soil. J. Appl. Sci. **5**(9), 1669–1673 (2005)

Ashok, P., Sureshkumar, M.: Experimental studies on concrete utilising red mud as a partial replacement of cement with hydrated lime. J. Mech. Civil Eng., 1–10 (2014)

Aziz, M.A., Paramasivam, P., Lee, S.L.: Prospects for natural fibre reinforced concretes in construction. Int. J. Cem. Compos. Lightweight Concrete **3**(2), 123–132 (1981)

Bagherpour, I., Choobbasti, A.J.: Stabilization of fine-grained soils by adding microsilica and lime or microsilica and cement. Electron. J. Geotech. Eng. **8**, 1–10 (2003)

Balamurugan, G., Perumal, P.: Use of quarry dust to replace sand in concrete–an experimental study. Int. J. Sci. Res. Publ. **3**(12) (2013)

Balwaik, S.A., Raut, S.: Utilization of waste paper pulp by partial replacement of cement in concrete. Int. J. Eng. Res. Appl. **1**(2), 300–309 (2011)

Bell, F.G.: Lime stabilization of clay minerals and soils. Eng. Geol. **42**(4), 223–237 (1996)

Bhuvaneshwari, S., Robinson, R., Gandhi, S.: Stabilization of expansive soils using fly ash. Fly Ash India **8**, 5.1–5.10 (2005)

Bohrn, G., Stampfer, K.: Untreated wood ash as a structural stabilizing material in forest roads. Croatian J. Forest Eng. **35**(1), 81–89 (2014)

Bolden, J.J.: Innovative uses of Recycled and Waste Materials in Construction Application. North Carolina Agricultural and Technical State University (2013)

Boraste, M.T.H., Sharma, V.K.: Investigation of laterite soil by waste glass for stabilization of road embankment (2014)

Bose, B.: Geo engineering properties of expansive soil stabilized with fly ash. Electron. J. Geotech. Eng. **17**, 1339–1353 (2012)

Cairns, R., Kenny, M.: The use of recycled rubber tyres in concrete. Used/Post-Consumer Tyres **3**, 135 (2004)

Chansoria, A., Yadav, R., Chansoria, A., Yadav, R.: Effect of quarry dust on engineering properties of black cotton soil. Int. J. **2**, 715–718 (2016)

Cheah, C.B., Ramli, M.: The implementation of wood waste ash as a partial cement replacement material in the production of structural grade concrete and mortar: an overview. Resour. Conserv. Recycl. **55**(7), 669–685 (2011)

Chong, W.K., Kumar, S., Haas, C.T., Beheiry, S.M., Coplen, L., Oey, M.: Understanding and interpreting baseline perceptions of sustainability in construction among civil engineers in the United States. J. Manag. Eng. **25**(3), 143–154 (2009)

Correia, A.G., Winter, M., Puppala, A.: A review of sustainable approaches in transport infrastructure geotechnics. Transp. Geotech. **7**, 21–28 (2016)

De Bock, L., Gonçalves, A.: Recycled asphalt pavement. Rilem Report **30**, 45 (2005)

Edil, T.B., Acosta, H.A., Benson, C.H.: Stabilizing soft fine-grained soils with fly ash. J. Mater. Civ. Eng. **18**(2), 283–924 (2006)

Erfen, Y.B., Yunus, K.N.B.M.: The appropriateness of egg shell as filler in hot mix asphalt (2015)

Ganesan, K., Rajagopal, K., Thangavel, K.: Rice husk ash blended cement: assessment of optimal level of replacement for strength and permeability properties of concrete. Constr. Build. Mater. **22**(8), 1675–1683 (2008)

Gayathri, V.: Utility of lime and red mud in clay soil stabilization (2016)

Goodarzi, A.R., Salimi, M.: Stabilization treatment of a dispersive clayey soil using granulated blast furnace slag and basic oxygen furnace slag. Appl. Clay Sci. **108**, 61–69 (2015)

Hanif, M., NurShafiq, M., Tajuddin, A., Azhar, S., Abdul Kadir, A., Madun, A., Azmi, M., Azim, M., Nordin, N.S.: Leachate characteristics of contaminated soil containing lead by stabilization/solidification technique (2014)

Hill, R.C., Bowen, P.A.: Sustainable construction: principles and a framework for attainment. Constr. Manag. Econ. **15**(3), 223–239 (1997)

Hua-dong, M., Liu, L.: Stability processing technology and application prospect of steel slag. Steelmaking **25**(6), 74 (2009)

Huang, B., Shu, X., Burdette, E.: Mechanical properties of concrete containing recycled asphalt pavements. Mag. Concr. Res. **58**(5), 313–320 (2006)

Hut, M.N.S.: The performance of eggshell powder as an additive concrete mixed, UMP (2014)

Ikara, I., Kundiri, A., Mohammed, A.: Effects of Waste Glass (WG) on the strength characteristics of cement stabilized expansive soil. Am. J. Eng. Res. **4**(11), 33–41 (2015)

James, J., Pandian, P.K.: Development of early strength of lime stabilized expansive soil: effect of red mud and egg shell ash. Acta Technica Corviniensis-Bulletin of Engineering **9**(1), 93 (2016)

James, J., Pandian, P.K., Switzer, A.: Egg shell ash as auxiliary addendum to lime stabilization of an expansive soil. J. Solid Waste Technol. Manag. **43**(1), 15–25 (2017)

Kalkan, E.: Utilization of red mud as a stabilization material for the preparation of clay liners. Eng. Geol. **87**(3), 220–229 (2006)

Kalkan, E.: Effects of silica fume on the geotechnical properties of fine-grained soils exposed to freeze and thaw. Cold Reg. Sci. Technol. **58**(3), 130–135 (2009)

Kalkan, E., Akbulut, S.: The positive effects of silica fume on the permeability, swelling pressure and compressive strength of natural clay liners. Eng. Geol. **73**(1), 145–156 (2004)

Khan, M.I., Siddique, R.: Utilization of silica fume in concrete: review of durability properties. Resour. Conserv. Recycl. **57**(Supplement C), 30–35 (2011)

Kharade, A.S., Suryavanshi, V.V., Gujar, B.S., Deshmukh, R.R.: Waste product 'Bagasse ash' from sugar industry can be used as stabilizing material for expansive soils. Int. J. Res. Eng. Technol. **3**(3), 506–512 (2014)

Kumar, U.A., Biradar, B.: Soft subgrade stabilization with quarry dust-an industrial waste. Int. J. Res. Eng. Technol. **3**(8), 1–4 (2014)

Maghool, F., Arulrajah, A., Du, Y.-J., Horpibulsuk, S., Chinkulkijniwat, A.: Environmental impacts of utilizing waste steel slag aggregates as recycled road construction materials. Clean Technol. Environ. Policy **19**(4), 949–958 (2017)

Mane, N., Rajashekhar, M.: Stabilization of black cotton soil by using red mud and sodium silicate (2017)

Manikandan, A., Moganraj, M.: Consolidation and rebound characteristics of expansive soil by using lime and bagasse ash. Int. J. Res. Eng. Technol. **3**(4), 403–411 (2014)

Mazloom, M., Ramezanianpour, A., Brooks, J.: Effect of silica fume on mechanical properties of high-strength concrete. Cement Concr. Compos. **26**(4), 347–357 (2004)

Mirzababaei, M., Miraftab, M., McMahon, P., Mohamed, M.: Undrained behaviour of clay reinforced with surplus carpet fibres. In: Second International Symposium on Fiber Recycling, Atlanta, Georgia, USA (2009)

Mounika, K., Narayana, B.S., Manohar, D., Vardhan, K.S.H.: Influence of sea shells powder on black cotton soil during stabilization. Int. J. Adv. Eng. Technol. **7**(5), 1476 (2014)

Nadgouda, K.: Coconut fibre reinforced concrete. In: Thirteenth IRF International Conference, 14th September (2014)

Najar, P.A.M., Nimje, M., Bagde, S.U., Pathak, V.S., Prajapati, S., Mukhopadhyay, J., Satpathy, B.K.: Development of light weight foamed bricks from red mud (2012)

Nalbantoglu, Z., Tawfiq, S.: Evaluation of the effectiveness of olive cake residue as an expansive soil stabilizer. Environ. Geol. **50**(6), 803–807 (2006)

Negi, C., Yadav, R., Singhai, A.: Effect of silica fume on engineering properties of black cotton soil. Int. J. Comput. Eng. Res. **3**(7), 1–10 (2013)

Obuzor, G., Kinuthia, J., Robinson, R.: Enhancing the durability of flooded low-capacity soils by utilizing lime-activated ground granulated blastfurnace slag (GGBS). Eng. Geol. **123**(3), 179–186 (2011)

Ochepo, J., Sadeeq, J., Salahudeen, A., Tijjani, S.: Effect of bagasse ash on lime stabilized lateritic soil. Jordan J. Civil Eng. **9**(2), 203–213 (2015)

Okonkwo, U., Odiong, I., Akpabio, E.: The effects of eggshell ash on strength properties of cement-stabilized lateritic. Int. J. Sustain. Constr. Eng. Technol. **3**(1) (2012)

Olufowobi, J., Ogundoju, A., Michael, B., Aderinlewo, O.: Clay soil stabilization using powdered glass. J. Eng. Sci. Technol. **9**(05), 541–558 (2014)

Oriola, F., Moses, G.: Groundnut shell ash stabilization of black cotton soil. Electron. J. Geotech. Eng. **15**, 415–428 (2010)

Osinubi, K., Ijimdiya, T.S., Nmadu, I.: Lime stabilization of black cotton soil using bagasse ash as admixture. Adv. Mater. Res. **62**, 3–10 (2009)

Othman, N.H., Bakar, B.H.A., Don, M., Johari, M.: Cockle shell ash replacement for cement and filler in concrete. Malaysian J. Civil Eng. **25**, 201–211 (2013)

Pliya, P., Cree, D.: Limestone derived eggshell powder as a replacement in Portland cement mortar. Constr. Build. Mater. **95**(Supplement C), 1–9 (2015)

Poh, H., Ghataora, G.S., Ghazireh, N.: Soil stabilization using basic oxygen steel slag fines. J. Mater. Civ. Eng. **18**(2), 229–240 (2006)

Puppala, A.J., Saride, S., Williammee, R.: Sustainable reuse of limestone quarry fines and RAP in pavement base/subbase layers. J. Mater. Civ. Eng. **24**(4), 418–429 (2011)

Qasrawi, H., Shalabi, F., Asi, I.: Use of low CaO unprocessed steel slag in concrete as fine aggregate. Constr. Build. Mater. **23**(2), 1118–1125 (2009)

Radhakrishnan, G., Kumar, M.A., Raju, G.: Swelling properties of expansive soils treated with chemicals and fly ash. Am. J. Eng. Res. **3**(4), 245–250 (2014)

Raji, S., Samuel, A.: Egg shell as a fine aggregate in concrete for sustainable construction. Int. J. Sci. Technol. Res. **4**(09), 1–13 (2015)

Rao, A.N., Chittaranjan, M.: Applications of agricultural and domestic wastes in geotechnical applications: an overview. J. Environ. Res. Dev. **5**(3) (2011)

Roy, A.: Soil stabilization using rice husk ash and cement. Int. J. Civil Eng. Res. **5**(1), 49–54 (2014)

Sabat, A.K.: A study on some geotechnical properties of lime stabilised expansive soil–quarry dust mixes. Int. J. Emerg. Trends Eng. Dev. **1**(2), 42–49 (2012)

Sabat, A.K., Bose, B.: Improvement in geotechnical properties of an expansive soil using fly ash–quarry dust mixes. Electron. J. Geotech. Eng. **18**, 3487–3500 (2013)

Sabat, A.K., Pati, S.: A review of literature on stabilization of expansive soil using solid wastes. Electron. J. Geotech. Eng. **19**, 6251–6267 (2014)

Sabat, A.K., Pradhan, A.: Fiber reinforced–fly ash stabilized expansive soil mixes as subgrade material in flexible pavement. Electron. J. Geotech. Eng. **19**, 5757–5770 (2014)

Sadeeq, J., Ochepo, J., Salahudeen, A., Tijjani, S.: Effect of bagasse ash on lime stabilized lateritic soil. Jordan J. Civil Eng. **9**(2) (2015)

Schettino, M.A.S., Holanda, J.N.F.: Characterization of sugarcane bagasse ash waste for its use in ceramic floor tile. Procedia Materials Science **8**(Supplement C), 190–196 (2015)

Seyyedalipour, S.F., Kebria, D.Y., Malidarreh, N., Norouznejad, G.: Study of utilization of pulp and paper industry wastes in production of concrete. Int. J. Eng. Res. Appl. **4**(1), 115–122 (2014)

Siddique, R.: Utilization of silica fume in concrete: review of hardened properties. Resour. Conserv. Recycl. **55**(11), 923–932 (2011)

Soosan, T., Sridharan, A., Jose, B.T., Abraham, B.: Utilization of quarry dust to improve the geotechnical properties of soils in highway construction (2005)

Srinivasan, R., Sathiya, K.: Experimental study on bagasse ash in concrete. Int. J. Serv. Learn. Eng. Humanitarian Eng. Soc. Entrepreneurship **5**(2), 60–66 (2010)

Surjandari, N.S., Djarwanti, N., Ukoi, N.U.: Enhancing the engineering properties of expansive soil using bagasse ash. J. Phys: Conf. Ser. **909**, 012068 (2017)

Taha, R., Al-Harthy, A., Al-Shamsi, K., Al-Zubeidi, M.: Cement stabilization of reclaimed asphalt pavement aggregate for road bases and subbases. J. Mater. Civ. Eng. **14**(3), 239–245 (2002)

Vaníček, I., Jirásko, D., Vaníček, M.: Added value of transportation geotechnics to the sustainability (design approach). Procedia Eng. **143**, 1417–1424 (2016)

Vaníček, I., Vaníček, M.: Modern earth structures of transport engineering view of sustainable construction. Procedia Eng. **57**(Supplement C), 77–82 (2013)

Wang, Y.: Utilization of recycled carpet waste fibers for reinforcement of concrete and soil. Polym. Plast. Technol. Eng. **38**(3), 533–546 (1999)

Wang, Y.: Utilization of recycled carpet waste fibers for reinforcement of concrete and soil. Recycl. Text. **238**(7), 213–224 (2006)

Wang, Y., Zureick, A.-H., Cho, B.-S., Scott, D.: Properties of fibre reinforced concrete using recycled fibres from carpet industrial waste. J. Mater. Sci. **29**(16), 4191–4199 (1994)

Yerramala, A.: Properties of concrete with eggshell powder as cement replacement. Indian Concr. J., 94–102 (2014)

Yi, H., Xu, G., Cheng, H., Wang, J., Wan, Y., Chen, H.: An overview of utilization of steel slag. Procedia Environ. Sci. **16**, 791–801 (2012)

Mechanical Characterization of the Adobe Material of the Archaeological Site of Chellah

S. Simou[1($\boxtimes$)], K. Baba[2], N. Akkouri[1], M. Lamrani[2], M. Tajayout[2], and A. Nounah[2]

[1] Civil Engineering and Environment Laboratory, High School of Technology, Sale, Civil Engineering, Water, Environment and Geosciences Centre (CICEEG), Mohammadia School of Engineering, Mohammed V University, Rabat, Morocco
sanaesimou@research.emi.ac.ma

[2] Civil Engineering and Environment Laboratory, High School of Technology, Sale, Mohammed V University, Rabat, Morocco

Abstract. Monuments and historical remains built of raw earth show that this material can persist for centuries. As with the cultural and historical heritage of the city of Rabat (Morocco), Chellah is one of the most historically significant monuments, this site being the melting pot of several civilizations that have played a major role, including the Marinid who left traces of Islamic architecture through the Medersa site built by the adobe construction technique.

This research involves the experimental and numerical analysis of the adobe material obtained from the Chellah site as well as the additive mixture in order to identify their mechanical behaviour. After the geotechnical identification of the material, a series of mechanical tests (compressive and bending tensile strengths) were applied to the adobe and the additive mixture. For numerical simulation, the finite element method was used to simulate the failure process of the material and different mixtures using the ANSYS software (ANSYS, 2019 1R). Numerical predictions are compared with experimental data. The methodology applied in this study provides promising results to better predict the mechanical properties of the mix used on the construction and rehabilitation of historic monuments in order to further reduce costly experimental tasks.

Keywords: Historical monument · Building materials · Adobe · Mechanical behaviour · Wood shaving

1 Introduction

In Morocco, the evaluation of historic buildings has always posed significant challenges due to the difficulties associated with the characterization of generally complex geometries, the variability in the properties of building materials and uncertainties about the actual state of damage to these constructions. This issue is even more complex when it comes to historical adobe masonry buildings, as earthen masonry has great variability and rapid deterioration over time if it is not properly maintained. In the context of the above, it was important to provide information in order to support intervention projects in historic centres. In this perspective, we intend to provide information about the mechanical and physical characterization of an ancient masonry,

© Springer Nature Switzerland AG 2020
H. Ameen et al. (Eds.): GeoMEast 2019, SUCI, pp. 118–130, 2020.
https://doi.org/10.1007/978-3-030-34199-2_8

through a case study of the adobe material taken from the archaeological site Chellah, which represents the remains of the first human occupation of the city of Rabat (Morocco), since antiquity (7th–6th century BC) (Terrisse 2011).

The adobe material is composed of adequate proportions of gravel, sand, silt, clay, and water, and it is produced by beating and compacting soil in a formwork (CRATerre 1979). In recent years, there has been occasional recourse to the rehabilitation and reinforcement of adobe buildings by some owners who are aware of its preservation and protection (CRATerre 2018). The rehabilitation option presents difficulties related to the lack of information on the properties and characteristics of the mechanical behaviour of adobe, indeed technical studies are needed to determine properties such as: modulus of elasticity, compressive, tensile and shear strength, and its composition (Varum et al. 2008).

The improvement of the soil using a stabilizer and an appropriate reinforcer material, natural or artificial, improves not only its durability but also its compressive strength, the latter increases considerably by adding fibres (Sharma et al. 2015, 2016). Different stabilizers produce different impacts on the soil's durability. The most commonly used natural stabilizers are jute, sisal, straw, rice husk, sugar cane bagasse (Alavéz-Ramírez et al. 2012), and soluble silanes or siloxanes, isocyanates, lime, cement, plaster, basalt pumice (Ghavami et al. 1999) are among the most frequently used synthetic stabilizers. In this context, samples of adobe taken from the Chellah site were mixed with different wood shaving contents. In the case, wood fibre materials have a thermal conductivity close to that of expanded polystyrene and glass wool (Koohestani et al. 2016). In addition, they are easy to use for both new construction and renovations. They also have other advantages such as excellent sound isolation due to their ability to absorb sound waves. Another major advantage of adding wood shaving to building materials is that they are inexpensive, available in large quantities, environmentally friendly and easy to process (Sudin and Swamy 2006; Taj and Munawar 2007; Onuaguluchi and Banthia 2016).

The experimental program consisted in a first step of physical characterization including an analysis of the granulometric distribution of the adobe material and a statistical analysis of the content of the wood fibres added to the mix of the adobe by volume, in the second step a study of uniaxial compression tests on cylindrical samples, and bending tensile tests on prismatic samples according to the three-point loading schema. These tests make it possible to evaluate the strength capacity of the specimen, and the evolution of rigidity and deformation with the increase in the content of wood shaving. In addition, an evaluation by the finite element method, in order to simulate the arbitrary propagation of cracks in adobe and wood shaving specimens. The results of the numerical modelling are compared with the experimental results. The methodology applied in this study provides promising results to better predict for the mechanical properties used in construction and to further reduce costly experimental work.

2   Experimental Program

2.1   Materials and Proportions of the Mixture

The adobe material used in this study comes from the Chellah archaeological site (latitude: 34.006686, longitude: −6.820532), it is a mixture of gravel and sand, clay and silt. Wood shaving are made of Scotch pine tree waste produced from wood-working activities Fig. 1a and have an irregular shape, with a particle size distribution between 2 mm and 30 mm. The apparent density of the shaving is about 46.8 kg/m^3.

Fig. 1. (a) Wood shaving (b) cylindric specimens with wood shaving

Two different mixtures were prepared: a reference mixture composed only of adobe soil and a mixture with wood shaving in different proportions: 10%, 20%, 30%, 40%, 50%, 60%, and 70% by volume of wood shaving. The water content on the mixtures was determined by the Proctor test. Table 1 summarizes the different mixing ratios and dry densities of samples with wood shaving. Two types of specimens were manufactured for this study Fig. 2.

Table 1. Proportion of the mixture and dry density of the samples

Specimens	Proportions of wood shavings (%)	Water content (%)	Dry density (kg.m^{-3})
S1	0	13.3	1883,0
S2	10	15.38	1699,7
S3	20	17	1589,2
S4	30	19	1720,9
S5	40	19.25	1679,5
S6	50	20.51	1761,2
S7	60	21.98	1659,9
S8	70	21.9	1559

Cylindrical specimens 160 mm in diameter and 320 mm high intended for compressive strength testing. To perform the bending tensile tests, rectangular prismatic specimens 4 × 4 × 16 cm^3 were manufactured.

Fig. 2. (a) Compressive test (b) Three-point bending test

All samples were left in the open air for four weeks, and then put in an oven to be dried at 60 °C for 24 h, and then the temperature was raised to 100 °C. The specimens were then stored in a controlled room at 20 °C and 50% relative humidity (RH) and were tested when they were in equilibrium with the environment about two weeks later.

2.2 Geotechnical Identification and Classification

Several laboratory tests are carried out, these tests are carried out with reference to French standards. The determination of the granulometric distribution indicates that the formulation of the adobe soil corresponds to a mixture of 9% gravel, 52.5% sand, 22.5% clay, with a significant fraction of silt about 16%.

The consistency study was conducted on the basis of the Atterberg limits, which are respectively: plastic limit W_p = 25.90% and liquid limit WL = 33.37%, and the plasticity index was calculated as the difference between the liquidity limit and the plasticity limit I_p = 7.47%. For the normal Proctor optimum, the water content (W_{opt}) and the maximum dry density $\gamma_{d\,max}$, are respectively 13.3%, and 1.87 g/cm^3. The VBS value is 1.33. The characteristics of the grain, as well as data from the documentation of the seismic regulations earth structures (RPCT 2013) indicate that this material is appropriate for the chosen purpose.

2.3 Compression Tests

Bui Quoc-Bao and his collaborators studied the compressive strength of the adobe in the perpendicular and parallel directions to the beds on prismatic samples, which gave results that are not very different (Quốc-Bảo 2008) to the other results of the simple compression of the other studies (Maniatidis and Walker 2008). The compressive strength of unstabilized soil increases from a few tenths of MPa for air-dried soil (0.5 to 1.5 MPa for cob; 1.0 to 2.5 MPa for adobe) to a few MPa for compacted soil (1 to 4 MPa) and compressed blocks (1 to 7 MPa) (Van Damme and Houben 2018). Several studies have concluded that there has been a significant improvement in compressive strength through the addition of stabilizing additives (Quốc-Bảo 2008; Calatan et al. 2016; González-López et al. 2018).

Samples were prepared to investigate the compressive strength and important characteristics of the adobe material (six samples) and the adobe material containing wood shaving as an additive. Thirteen cylindrical (160 × 320 mm) samples were subjected to compression tests Fig. 2. The tests were carried out according to (EN 772-1: 2011+A1 2015).

The compressive strength results of the adobe material are presented in Table 2, the average compressive strength value is approximately 0,54 MPa. Figure 3 shows the compressive strength results of the adobe material containing wood shaving for all tested proportions (10%, 20%, 30%, 40%, 50%, 60%, and 70%) by volume of wood shaving, the maximum strength is 1.68 MPa which corresponds to the proportion 50%.

Table 2. Compressive strength obtained from the tests on cylindric samples of adobe material

Samples	1	2	3	4	5	6	Moyenne	Ecart-type
Stress (MPa)	0.48	0.50	0.53	0.50	0.58	0.63	**0.54**	0.054

By comparing the results obtained during the first compression test on the adobe material Table 2 with those obtained on the adobe with the wood shaving Fig. 3, it is quite clear that there is a significant improvement in the compressive strength.

2.4 Bending Tensile Tests

As the bending tests are very much influenced by the experimental framework, a lot of research has been developed previously either on concrete or on earthen materials. The experimental program of this study was designed to evaluate the mechanical behaviour of the adobe and reinforced adobe specimens using the bending tensile test according to the standards (EN 196-1 2016), in order to evaluate the rigidity and load-bearing capacity of this material.

The mechanical tests were carried out at a constant deformation rate of 0.5 mm/min, each sample tested was of a prismatic rectangular shape with dimensions 4 × 4 × 16 mm Fig. 1b, six samples of the adobe material and seven samples of the adobe reinforced with different proportions of wood shaving (10%, 20%, 30%, 40%, 50%, 60%, and 70%), in total, thirteen samples were tested. Geometry and load are illustrated in Fig. 5.

The average tensile strength found for the samples of the adobe material is about 0.22 MPa, and for the samples reinforced the maximum strength is 4.85 MPa which corresponds to the proportion 50% by volume of wood shaving, and the average strength is 2.63 MPa Fig. 5.

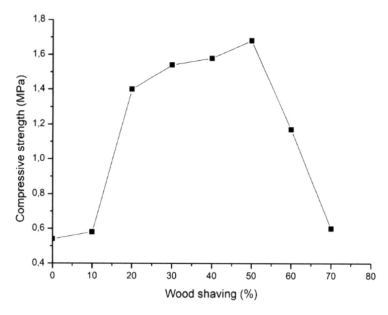

Fig. 3. Compressive strengths of adobe material with different proportions of wood shaving

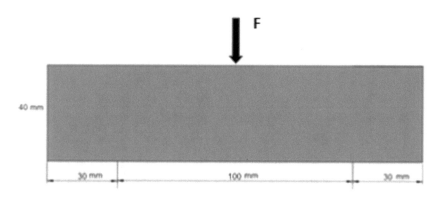

Fig. 4. Geometric and load characteristics

Figure 6, has been obtained experimentally for samples of the adobe material (S1) it is around 73 MPa and the reinforced one (S2-S3-S4-S5-S6-S7-S8) is of an average value of 174.35 MPa. The mixture of adobe material and wood shaving has a higher modulus of elasticity than the modulus of the unreinforced material.

By comparing the results obtained from the bending tensile tests Fig. 5 with those obtained from the compression tests Fig. 3, it is quite obvious that the compressive strength is more than the bending strength values for unreinforced adobe material, the ratio between bending and compressive strength for the same material is on average 40.7%. On the other hand, we can see that the improvement in tensile strength for reinforced adobe is more significant than the compressive strength Fig. 7.

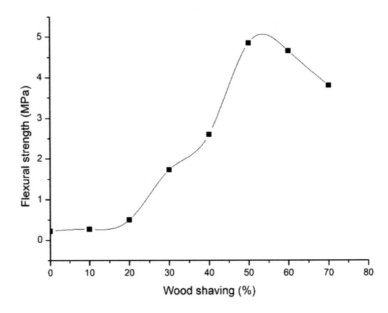

Fig. 5. Flexural strength of adobe material with different proportions of wood shaving

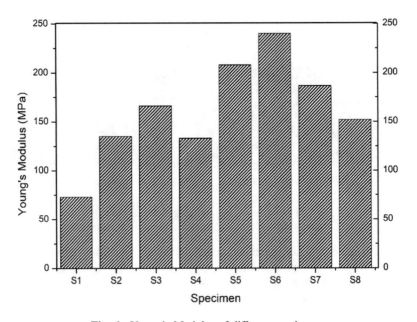

Fig. 6. Young's Modulus of different specimens

2.4.1 Numerical Simulation

Advanced numerical analysis methods have been widely used to simulate conventional masonry structures such as concrete, stone and earth structures (Mohebkhah and

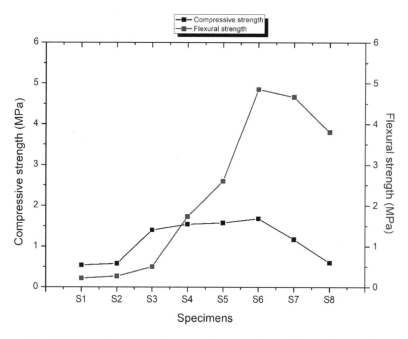

Fig. 7. Comparison curve of compressive strengths and flexural strengths

Chegeni 2012; Bui et al. 2017). On the other hand, numerical research and the application of its tools in adobe masonry structures has not been rigorous enough (Illampas et al. 2014) and less systematic, it has focused on the development of continuous finite element models (Tarque et al. 2010) and discrete elements (Cao and Watanabe 2004) using existing constituent laws. In the case of rammed earth construction, however, experimental studies to analyze the structural behaviour of the construction were conducted in combination with non-linear numerical analysis to examine the use of finite element models for the simulation of laterally loaded, non-reinforced rammed earth masonry structures (Illampas et al. 2014). Studies have also been carried out to characterize behaviour experimentally and numerically using the discrete element method on a real and small scale to validate the ability of the MED to model adobe houses (Daudon et al. 2014).

The present document aims to combine experimental laboratory tests and numerical analysis consisting of pre-treatment, resolution of the FE model and subsequent post-treatment of the bending tensile specimens of the adobe material reinforced by wood shaving, using the static structural module of the ANSYS Workbench software (ANSYS 2019 1R).

2.5 Simulations of Compressive and Three-Point Bending Tests

A 3D analysis was proposed to model a simply supported adobe-wood shaving specimens loaded. Some modeling parameters are required to simulate the finite element model of the specimens of the adobe-wood shaving for test of three-point bending

(256 elements). The numerical values of the parameters can be defined according to experiments and literature if they are not known Table 3. Statically in three-point bending, the boundary conditions, the geometry, and the loading are given in Fig. 4.

Table 3. Mechanical parameters of the simulation

Materials	Density ρ (Kg/m^{-3})	Flexural strength (MPa)	Compressive strength (MPa)	Young's Modulus (MPa)
Adobe material	1883	0.22	0.54	73
Adobe-wood shaving	1761.2	4.85	1.68	240

Figure 8 shows a contour plot of the total deformation, we have a minimum of 0.041 mm and a maximum of 0.372 mm. Figures 9 and 10 show strain and stress plots along the x axis that explain the tensile and compressive stresses. If we compare the Flexural stress of our simulation to the experimental value, we see in Fig. 11 that we have a maximum stress of 4.61 MPa, which compares quite well with the experimental value of 4.85 MPa Fig. 5.

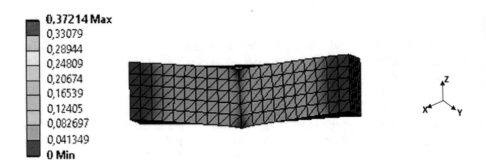

Fig. 8. Contour plot of the total displacements (mm)

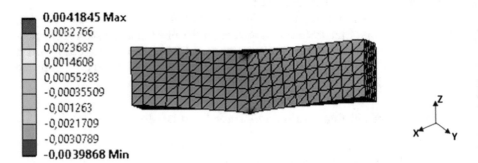

Fig. 9. Elastic normal strain along x (mm/mm)

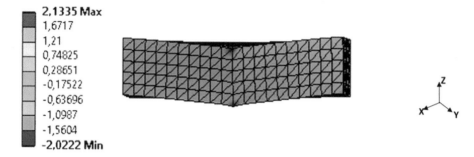

2,1335 Max
1,6717
1,21
0,74825
0,28651
-0,17522
-0,63696
-1,0987
-1,5604
-2,0222 Min

Fig. 10. Elastic normal stress along x (MPa)

4,6113 Max
4,1
3,5888
3,0775
2,5662
2,0549
1,5436
1,0323
0,52104
0,0097534 Min

Fig. 11. Flexural strength (MPa)

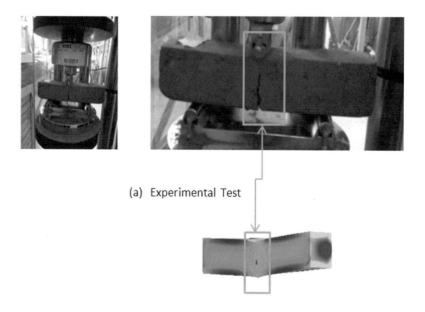

(a) Experimental Test

(b) Numerical test

Fig. 12. Mode of rupture of the adobe-wood specimen subjected to bending test

2.5.1   Comparison Between Modelling and Experimental Results

The failure mode is related to the physical characteristics of each specimen. But several samples often fail in the tensile zone, Fig. 12. During the bending test, the adobe-Wood shaving specimen was subjected to tensile stresses on one side of the neutral axis, while on the other side of the neutral axis the specimen was subjected to compressive stresses as shown in Fig. 11. The failure initiated from the tensile zone and propagated upwards through the compression zone.

A comparison between the failure of the presented model and that calculated experimentally showed in Fig. 12. It can be seen that the model predicted the damaged area in the lower fibers matching the tensile zone, and the experimental failure actually occurred at the same position.

3   Conclusions

This research paper is in the interest of providing information on the mechanical and physical characterization and reinforcement techniques of the adobe material of the archaeological site Chellah (Rabat-Morocco). A mixture has been studied; the adobe material with wood shaving tested in different proportions, the experimental tests are carried out on different specimens, tested in compression and tensile bending, and afterwards the numerical test was performed using a finite element analysis by (ANSYS 2019 1R) software.

Experimental tests show that mechanical properties such as tensile strength and modulus of elasticity can be identified by bending tensile tests, and that the adobe material resists both in compression and tensile stress, as well as the ratio between bending strength and compressive strength is 40%. The addition of wood shaving has improved the performance of the mix made by adobe and wood shavings in compression and tensile strength, and that the addition of wood fibers has a significant improvement in tensile strength as well as in compression.

The results obtained by applying the finite element analysis model used by ANSYS to evaluate the mechanical behaviour of the adobe and wood shaving specimens are in accordance with the values obtained by the experimental tests. This result will allow the use of the FEA method for the analysis of specimens of the adobe material.

The results obtained are required for the characterization of the adobe material used in the construction of the Chellah archaeological and historical site (Rabat-Morocco), and constitute important reference values to be taken into consideration in the restoration and rehabilitation of the site's buildings and in the calibration of numerical models.

References

Alavéz-Ramírez, R., Montes-García, P., Martínez-Reyes, J., et al.: The use of sugarcane bagasse ash and lime to improve the durability and mechanical properties of compacted soil blocks. Constr. Build. Mater. **34**, 296–305 (2012). https://doi.org/10.1016/j.conbuildmat.2012.02.072

Bui, T.T., Limam, A., Sarhosis, V., Hjiaj, M.: Discrete element modelling of the in-plane and out-of-plane behaviour of dry-joint masonry wall constructions. Eng. Struct. **136**, 277–294 (2017). https://doi.org/10.1016/j.engstruct.2017.01.020

Calatan, G., Hegyi, A., Dico, C., Mircea, C.: Determining the optimum addition of vegetable materials in adobe bricks. Procedia Technol. **22**, 259–265 (2016). https://doi.org/10.1016/j.protcy.2016.01.077

Cao, Z., Watanabe, H.: Earthquake response predication and retrofitting techniques of adobe structures, no. 12 (2004)

CRATerre: Construire en Terre normes CRAterre (1979)

CRATerre: Réhabilitation et valorisation du bâti en pisé (2018)

Daudon, D., Sieffert, Y., Albarracín, O., et al.: Adobe construction modeling by discrete element method: first methodological steps. Procedia Econ. Finan. **18**, 247–254 (2014). https://doi.org/10.1016/S2212-5671(14)00937-X

EN 196-1: Methods of testing cement - Part 1: determination of strength (2016). https://standards.cen.eu. Accessed 26 June 2019

EN 772-1:2011+A1: Methods of test for masonry units - Part 1: determination of compressive strength (2015)

Ghavami, K., Toledo Filho, R.D., Barbosa, N.P.: Behaviour of composite soil reinforced with natural fibres. Cement Concr. Compos. **21**, 39–48 (1999). https://doi.org/10.1016/S0958-9465(98)00033-X

González-López, J.R., Juárez-Alvarado, C.A., Ayub-Francis, B., Mendoza-Rangel, J.M.: Compaction effect on the compressive strength and durability of stabilized earth blocks. Constr. Build. Mater. **163**, 179–188 (2018). https://doi.org/10.1016/j.conbuildmat.2017.12.074

Illampas, R., Charmpis, D.C., Ioannou, I.: Laboratory testing and finite element simulation of the structural response of an adobe masonry building under horizontal loading. Eng. Struct. **80**, 362–376 (2014). https://doi.org/10.1016/j.engstruct.2014.09.008

Koohestani, B., Koubaa, A., Belem, T., et al.: Experimental investigation of mechanical and microstructural properties of cemented paste backfill containing maple-wood filler. Constr. Build. Mater. **121**, 222–228 (2016). https://doi.org/10.1016/j.conbuildmat.2016.05.118

Mohebkhah, A., Chegeni, B.: Local–global interactive buckling of built-up I-beam sections. Thin-Walled Struct. **56**, 33–37 (2012). https://doi.org/10.1016/j.tws.2012.03.018

Onuaguluchi, O., Banthia, N.: Plant-based natural fibre reinforced cement composites: a review. Cement Concr. Compos. **68**, 96–108 (2016). https://doi.org/10.1016/j.cemconcomp.2016.02.014

Quốc-Bảo, B.: Stabilité des structures en pisé: Durabilité, caractéristiques mécaniques. Thèse, L'Institut National des Sciences Appliquees de Lyon (2008)

RPCT: Decret n° 2-12-666 du reglement parasismique pour les constructions en terre et instituant le Comité national des constructions en terre (2013)

Sharma, V., Marwaha, B.M., Vinayak, H.K.: Enhancing durability of adobe by natural reinforcement for propagating sustainable mud housing. Int. J. Sustain. Built Environ. **5**, 141–155 (2016). https://doi.org/10.1016/j.ijsbe.2016.03.004

Sharma, V., Vinayak, H.K., Marwaha, B.M.: Enhancing compressive strength of soil using natural fibers. Constr. Build. Mater. **93**, 943–949 (2015). https://doi.org/10.1016/j.conbuildmat.2015.05.065

Sudin, R., Swamy, N.: Bamboo and wood fibre cement composites for sustainable infrastructure regeneration. J. Mater. Sci. **41**, 6917–6924 (2006). https://doi.org/10.1007/s10853-006-0224-3

Taj, S., Munawar, M.A.: Natural fiber-reinforced polymer composite, no. 17 (2007)

Tarque, N., Camata, G., Espacone, E., et al.: Numerical modelling of in-plane behaviour of adobe walls, no. 12 (2010)

Terrisse, M.: Les musées de sites archéologiques appréhendés en tant que vecteurs de développement local à travers trois études de cas préfigurant la mise en valeur opérationnelle du site de Chellah. Thèse, Université de Maine, le Mans (2011)

Van Damme, H., Houben, H.: Earth concrete. Stabilization revisited. Cem. Concr. Res. **114**, 90–102 (2018). https://doi.org/10.1016/j.cemconres.2017.02.035

Varum, H., Costa, A., Pereira, H., et al.: Caracterização do comportamento estrutural de paredes de alvenaria de adobe, no. 10 (2008)

The Impact of Recycled Plastic Waste in Morocco on Bitumen Physical and Rheological Properties

Nacer Akkouri[1(✉)], Khadija Baba[2], Sana Simou[1],
Nassereddin Alanssari[3], and Abderrahman Nounah[2]

[1] Civil Engineering and Environment Laboratory (LGCE), Civil Engineering,
Water, Environment and Geosciences Centre (CICEEG), Mohammadia
Engineering School, Mohammed V University, Rabat, Morocco
akkouri.nacer@gmail.com
[2] Civil Engineering and Environment Laboratory (LGCE), School
of Technology-Salé, Mohammed V University, Rabat, Morocco
[3] Technical Centre for Plastics and Rubber (CTPC), Casablanca, Morocco

Abstract. Morocco is getting into the circular economy and wants to increase its overall recycling rate from 5% to more than 20%. Today we find that plastic waste production reaches the level of 1 million tons per year from which only 20 to 30% is recovered (Moroccan Federation of Plastics Processing 2016). In addition, we have the road sector in Morocco, where huge quantities of pure bitumen used increasingly, which affects greatly the economic balance of the country.

Therefore, this work enables us to study the impact of the addition of recycled plastic materials into the composition of bitumen in order to reduce the quantities of pure bitumen consumed and to recover plastic waste while improving the characteristics of bitumen used in road construction.

At first, the focus has been oriented to determine the abundance of recycled thermoplastic waste in Morocco, taking into account the needs of the existing plastics industries before analyzing the effect of adding different rates of recycled waste, namely Polypropylene (PP), Low Density Polyethylene (LDPE) and Polystyrene (PS), to a conventional 35/50 quality bitumen.

Basic rheological parameters such as penetration, softening point, elastic recovery and ageing (RTFOT) helped to determine the changes caused by each rate of plastic addition to the pure bitumen.

Tests showed that the penetrability of modified bitumen decreases, and its softening point as well as its elastic recovery increases, so as a result, the thermal sensibility and the aging rate of the new bitumen mixture decreases.

Keywords: Polymers · Recycled plastic waste · Modified bitumen · Bitumen

1 Introduction

Morocco suffers from two main problems that are the management of municipal solid waste, particularly used plastics because of their non-biodegradable characteristics and the premature deterioration of many road pavements due to excessive traffic.

© Springer Nature Switzerland AG 2020
H. Ameen et al. (Eds.): GeoMEast 2019, SUCI, pp. 131–145, 2020.
https://doi.org/10.1007/978-3-030-34199-2_9

For the plastic wastes on one hand, about 7 million tons of solid waste are consumed each year in Morocco, including 572,000 tons of plastic waste, which is expected to reach 12 million tons by 2020 (French Chamber of Commerce and Industry of Morocco (CFCIM) 2014). The plastic waste structure is composed of one or more high molecular weight organic polymers, which is solid in its finished state and can also flow under a specific state, causing pollution and environmental problems (Abdullah et al. 2017a, b).

On the other hand we have the premature deterioration of many road pavements in Morocco due to excessive traffic, as we know roads provide access to almost everywhere, they are the most popular transportation systems used worldwide for the transportation of passengers and freight (Köfteci et al. 2014). In the last few decades, Morocco has given great importance to road infrastructure, given its impact on the country's economy, social and touristic development. In 2018, the total length of the paved road network in Morocco has reached 44 180 km (Ministry of Equipment, Transport, Logistics and Water, Morocco 2019). However, the increase in transport demand is also leading to a rapid deterioration of road pavements, which represents a constant worry for the country officials.

There for to help overcome or mitigate these problems, it is necessary to increase the rate of recycled plastic waste and improve the properties of the binder used in road construction. To this end, and according to Morocco's new approach to circular economy, we aim in this paper to reuse local plastic waste in road construction in order to ensure effective solid waste management and improve the rheological properties of the bitumen binders by adding recycled plastic materials.

In the pavement field, bituminous binders play a key role determining the performance of bituminous asphalt (Yao et al. 2018). As known the factors responsible of asphalt pavement failure are higher traffic volume, and higher axle loads, combined with significant variations in temperatures (Carreau et al. 2000). Indeed, a large area of Morocco has a humid tropical climate with significant temperature variations, that lead to a phenomenon known as rutting that reduces considerably the pavement life (Carreau et al. 2000; Tayfur et al. 2007). The high demand for bituminous binders has impacted the price and quality of binders, which has encouraged research to discover new binders (Tsantilis et al. 2019). A lot of work has been presented in the literature to focus on the influence of bituminous binders on rutting behavior of asphalt mixtures (Alataş et al. 2012; John and David 2003; Isacsson and Zeng 1997).

And since packaging is the most important sector of the plastics processing market in Morocco the focus will be oriented to the under recycled packaging materials that are made mostly of polymers such as; Polypropylene (PP), Low Density Polyethylene (LDPE) and Polystyrene (PS). These materials are named thermoplastics (Plastics Europe 2017; Singh et al. 2017). The investigations have shown that, thermoplastics increase the stiffness at high temperatures and cracking resistance at low temperatures, and longer fatigue life of bitumen (Alataş and Yilmaz 2013; Zhu et al. 2014). In addition to the benefits reported, the researchers also mentioned some problems, such as the low compatibility between polymer modifiers and bitumen, which is monitored by their respective properties (Wang et al. 2010; Polacco et al. 2015).

2   Materials and Methods

2.1   Materials

The 35/50 penetration grade bitumen used in this study is manufactured by BITUMA, its physical properties are synthetized in Table 1. The recycled plastic waste collected from the recyclers in morocco is used as bitumen modifier. The plastic waste used are separated into 3 categories, namely Polypropylene (PP), Low Density Polyethylene (LDPE) and Polystyrene (PS), for which we conducted Physio-chemical tests that allowed us to determine the basic characteristics of the recycled material mixture corresponding to each category. its Physio-chemical properties are summarized in Table 2. In order to be correctly incorporated into the bitumen, the plastic was crushed, extruded and granulated with a cylindrical shape of about 3–4 mm long to obtain an optimal size to feed the bitumen mixer Fig. 1(a).

Table 1.   Main properties of bitumen

Properties	Test method	Results
Penetration, T = 25 °C (10–1 mm)	NM EN 1426	46,6
R&B Softening point (°C)	NM EN 1427	52
Density (g/cm^3)	NF EN ISO 3838	1,0501
Ductility at 25 °C (cm)	NF T66-006	>150

2.2   Preparation of Recycled Plastic Modified Bitumen Samples

For the preparation of Recycled Plastic Modified Bitumen (RPMB), we used melt-blending technique. The pure bitumen (3000 g) was heated in oven until its fluid condition was reached at 170 °C (NM EN 14023) then the polymer was slowly added Fig. 1(b). The speed of the mixer was kept between 5000 rpm and 10000 rpm, and temperature, between 170 °C and 180 °C. The concentration of PP, LDPE and PS were taken as 3; 5 and 6% by weight of the bitumen. It was observed that over 6% of the plastic, the modified binder becomes visually heterogeneous and the modification process becomes difficult to achieve. The mixing was continuous for 40 min-1 h to produce homogenous mixtures. The recycled plastic modified bitumen (RPMB) was then sealed in containers and stored for further testing.

2.3   Test Methods

The ring and ball (R&B), and penetration (Pen25) test were performed based on the NM EN 1427 and NM EN 1426 standard respectively, to evaluate the softening point and hardness of bitumen before and after modification. Then to determine the elastic

Table 2. Main properties of waste plastic

Properties	Test method	Polypropylene (PP)	Low Density Polyethylene (LDPE)	High impact polystyrene (PS)
Stress at break (MPa)	ISO 527-1	28,060	27,860	29,620
Elongation at break (%)		3,300	5,480	1,840
Young's modulus (MPa)		894,000	685,600	1392,000
Hardness Shore D	NF ISO 48-5	63,400	67,480	72,700
Density (g/cm^3)	ISO 1183 method A	0,879	0,870	1,061
Tf (°C)[a]	(DSC machine) ISO 11357	163	125	159
Tg (°C)[b]		–	–	99

[a]The melting temperature, [b]The glass transition temperature.

recovery of bituminous binders, we used a ductilimeter at the test temperature according to the NM EN 13398. Regarding the repeatability of the tests, the three results of the penetration test were kept under 2 dmm and the difference between two results of R&B and elastic recovery test were kept under 1.5 °C and 4% respectively.

The resistance to hardening under the influence of heat and air of the bitumen was measured using the Rolling Thin Film Oven Test (RTFOT) according to the NM EN 12607-1. In addition, sensitivity of the RPMBs has been calculated in terms of penetration index (PI). PI values were calculated by using the results obtained from penetration and softening point tests (John and David 2003) to calculate the PI values, the Eq. (1) was used according to the NM EN 12591.

$$PI = \frac{(1952 - 500 \times Log(Pen_{25}) - 20\,SP}{50 \times Log(Pen_{25}) - SP - 120} \tag{1}$$

The FTIR spectroscopy was conducted to monitor the effect of recycled plastic on the chemical structure of the pure bitumen. The FTIR spectra have been measured with Perkin Elmer Spectrum Two, spectrometer (with a resolution of 0.5 cm^{-1}) following the ATR procedure and spectrum readings were conducted between 400 and 4000 cm^{-1} intervals.

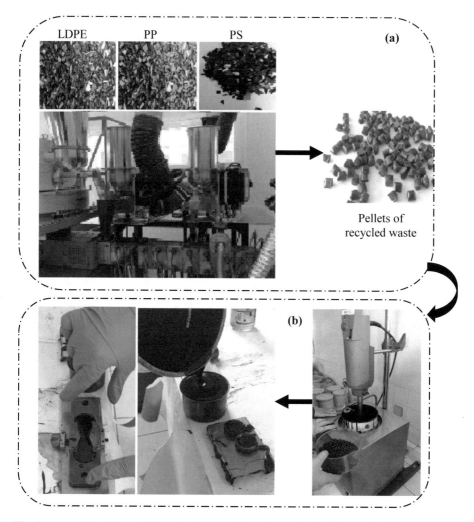

Fig. 1. (a) LDPE, PP and PS crushed, extruded and granulated with a cylindrical shape (3–4 mm); (b) melt-blending technique and testing of RPMBs

3 Results and Discussion

3.1 Penetration Results

As can be seen from the Fig. 2, the penetrability of the modified bitumen decreases with the addition of recycled plastic waste, from 46.6 for the pure bitumen to 25.8, 16.33 and 33.3 respectively for the RPMB at 6% of recycled PP, LDPE and PS. These results show the improvement that the addition of recycled plastic waste brings to the hardness of pure bitumen, which confirms the tendency of polymers to influence mainly the hardness of bitumen (Hunter et al. 2015). This hardness is explained by the swelling of the polyolefin components of the recycled plastic waste added into bitumen due to their absorbance of the light components of pure bitumen inducing the formation of a biphasic structure with a polyolefin phase in the bitumen matrix (Habib et al. 2010; Airey 2011; Zhu et al. 2014). Thus, modified bitumen becomes harder with the addition of recycled plastic waste, despite their insolubility, the polymer has continuously spreaded through the bitumen matrix and the properties obtained can probably be attributed to forming rigid network structure resisting deformation (Polacco et al. 2005, 2015).

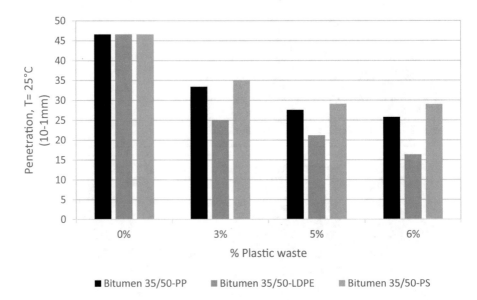

Fig. 2. Penetration results for PP, LDPE, PS and Bitumen

3.2 Softening Point Results

The results in the Fig. 3 shows that the softening point increased slightly with the addition of the recycled plastic waste ranging between 52 for pure bitumen to 59 for RPMB at 6% in recycled PP and PS except for the RPMB at 5% of recycled LDPE binder where the softening point increased considerably to reach 82. These results can be interpreted positively especially in enhancing the performance characteristic of the RPMB in term of rutting (Blazejowski and Dolzycki 2014).

3.3 Penetration Index Results

The penetration index PI can help determine the temperature susceptibility of the bitumen according to Pfeiffer and Van Doormaal (John and David 2003), this index shows the deviation of the binder behavior from Newtonien to Non-Newtonien. According to literature, the penetration index of bitumen ranges between +1 and −1 for road construction. When PI goes under −2 this shows a Newtonian behavior with more

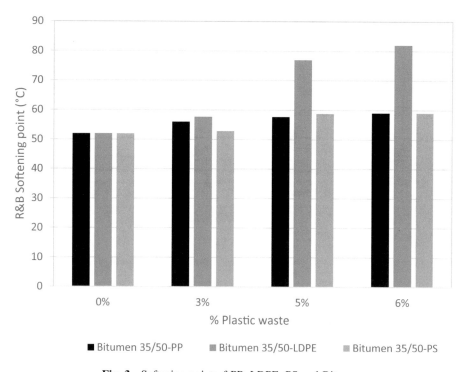

Fig. 3. Softening point of PP, LDPE, PS and Bitumen

temperature susceptibility, and when PI is greater than +2, the bitumen binder is less brittle with high elastic properties and with more susceptibility to low temperature (John and David 2003; Habib et al. 2010).

As showed in the Fig. 4 at the addition of 3% of recycled plastic waste the PI increased slightly for the PP and the LDPE except for the PS that decreased without breaking the −1 barrier. At the addition of 5% and above of the recycled plastic waste RPMB with LDPE, PI increased considerably to reach and exceed +2, which indicates that RPMB at ≥5% LDPE is brittle with high elastic properties and more susceptible to low temperature. For the RPMB with >5% PS and >5% PP we find that the PI increased but still at the pure bitumen PI interval.

3.4 Elastic Recovery Results

The elastic recovery is an interesting method for the assessment if the addition of recycled plastic waste to pure bitumen provides significant elastomeric characteristics for the binder (Hunter et al. 2015), but according to the test results in the Fig. 5, there is not considerable improvement for the RPMB at 3% recycled plastic waste. The elastic recovery even decreased afterwards for the RPMB with PP and PS, except for the case of RPMB with LDPE at 5% and 6%, which the high PI showed in the Fig. 4 explains the considerable increase (Habib et al. 2010), but these two mixtures reached the limit of rupture prematurely, which means that the binder's ductility decreased.

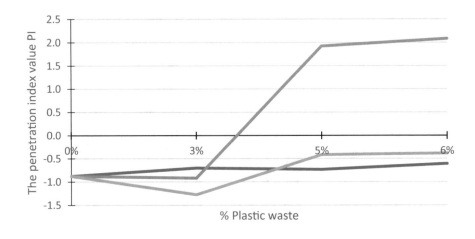

Fig. 4. Penetration index of PP, LDPE, PS and Bitumen

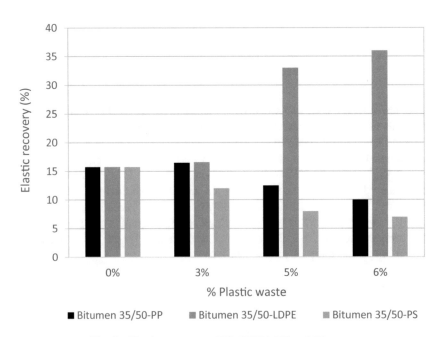

Fig. 5. Elastic recovery of PP, LDPE, PS and Bitumen

Since bitumen PI ranges between +1 and −1 for road construction (Habib et al. 2010), and according to PI test results for the RPMB binder we find that the mixtures fulfilling this condition with the highest PI are RPMB at 3% LDPE, RPMB at 5% PS and RPMB at 6% PP Table 3. Considering that the classification of bitumen depends on penetrability and softening point according to the NM EN 12591, we can say that the modification of pure bitumen 35/50 with recycled plastic waste upgraded the bitumen mixture class to 20/30 as summarized in Table 4.

Table 3. Penetration index of RPMBs at 3% LDPE, 5% PS, 6% PP

Test	RPMB at 3% PEBD	RPMB at 5% PS	RPMB at 6% PP
PI	−0,9	−0,4	−0,6

For the rest of this study the focus will be on the analysis of the mixtures chosen below.

3.5 Infrared Spectrometry Results

Bitumen chemical structure is a complex mixture of hydrocarbons containing different chemical compounds of relatively high molecular weight (Asphalt Institute and European Bitumen Association 2015). Thus, a complete analysis of bitumen in correlation with the rheological properties would be extremely laborious if not impossible (Hunter et al. 2015)

Following the IRTF analysis of pure bitumen, which serves as a reference, we observe that the peaks at 2850 and 2919 cm^{-1} are related to stretching vibrations characteristic of C-H bonds in aliphatic chains. Also we observe the appearance of a band centered at 1600 cm^{-1} characterizing the aromatic C=C contribution, while the peaks that appear at 1456 and 1375 cm^{-1} are due respectively to asymmetric deformations C-H in the CH_2 and CH_3 and symmetric deformations C-H in the CH3 vibrations (Brasileiro et al. 2019). We also note the existence of a 1700 cm^{-1} centred band characterizing the C=O carbonyls as well as another 1030 cm^{-1} centred band characterizing the S=O sulfoxides and finally a wide band between 3000 and 3600 cm^{-1} characterizing O-H hydroxyls. For the peaks that have been found in the 722–863 cm^{-1} region, they were associated with the C-H vibrations of the benzene ring.

In the FTIR spectra of the RPMB at 6% PP Fig. 6 and RPMB at 3% LDPE Fig. 7, we observe, in comparison with the pure bitumen spectra, a decrease in the peaks at 1700 cm^{-1}, 1030 cm^{-1} and in the band between 3000 and 3600 cm^{-1} that corresponds respectively to carbonyl, sulfoxide and hydroxyl. This variation indicates a decrease in the oxidation rate of the mixture, which is explained by the antioxidants present in the recycled plastic waste added to the pure bitumen. However, the FTIR analysis of the RPMB at 5% PS Fig. 8, point out that there is a small increase in carbonyl, sulfoxide and hydroxyl peaks, which indicate the presence of an ageing process during the preparation of the mixture that could be related to an initial oxidation of the recycled PS that favored the mixture oxidation.

The FTIR spectra of the pure bitumen and those of the three RPMB mixtures that modified the basic bitumen class, illustrate that there is not a big difference between the positions of the absorption peaks in the main absorption band for all spectra, which means that no new functional group were formed. Therefore, it can be deduced that this

Table 4. Penetration and softening point of RPMBs at 3% LDPE, 5% PS, 6% PP

Test	RPMB at 3% LDPE	RPMB at 5% PS	RPMB at 6% PP	Bitumen Class 20/30 (NM EN 12591)
Penetration, T = 25 °C (10–1 mm)	57,7	58,8	59	55–63
R&B Softening point (°C)	25	29,1	25,8	20–30

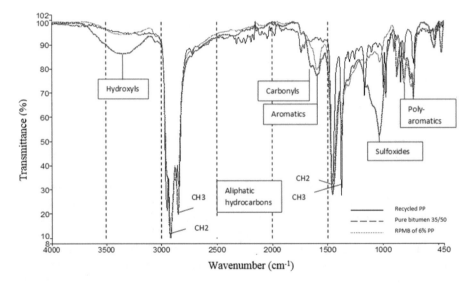

Fig. 6. FTIR spectra of recycled PP, pure bitumen 35/50 and RPMB-PP6% (Peak labelling according to Van den Bergh 2011; Weigel and Stephan 2017; Brasileiro et al. 2019)

enhancement in the peaks of the original structure is a result of a successful blending of the recycled plastic waste into the pure bitumen (Appiah et al. 2017; Nouali et al. 2019).

3.6   Bitumen Ageing Results

Mainly the aging process of a bitumen mixture consists on the loss of its most volatile components along with an oxidation process that leads to modifications in the mixture physical and chemical properties (Tauste et al. 2018).

After analyzing the conventional tests (Penetration and softening point), that have been conducted on the pure bitumen and the RPMB before and after RTFOT, we found that the three RPMB mixtures showed after RTFOT a less important increase in the softening point SP temperature. The Fig. 9 point out a variation of the SP by 6.3 °C, 1.2 °C and 4.4 °C respectively for the RPMB at 3% LDPE, 5% PS and 6% PP, which did not exceed the 8 °C recommended by the European standard NF EN 12591 to stay at level 1 severity. For the after RTFOT penetrability, we observed a less significant

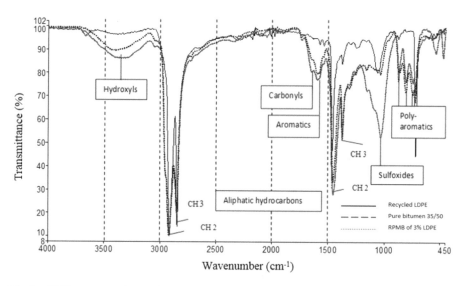

Fig. 7. FTIR spectra of recycled LDPE, pure bitumen 35/50 and RPMB-LDPE 3% (Peak labelling according to Van den Bergh 2011; Weigel and Stephan 2017; Brasileiro et al. 2019)

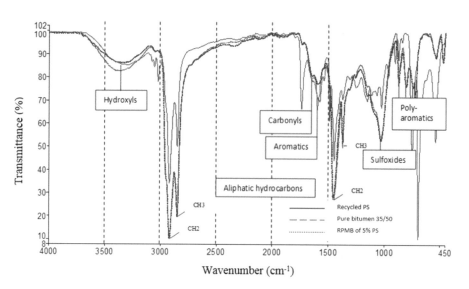

Fig. 8. FTIR spectra of recycled PS, pure bitumen 35/50 and RPMB-PS 5% (Peak labelling according to Van den Bergh 2011; Weigel and Stephan 2017; Brasileiro et al. 2019)

decrease for the three RPMB mixtures. Then the retained penetrability RP was calculated according to the Eq. (2). Figure 10 shows that the RP of the three mixtures are higher than that of the pure bitumen.

$$R\% = \frac{(P_{AGED})}{(P_{UN-AGED})} \times 100 \qquad (2)$$

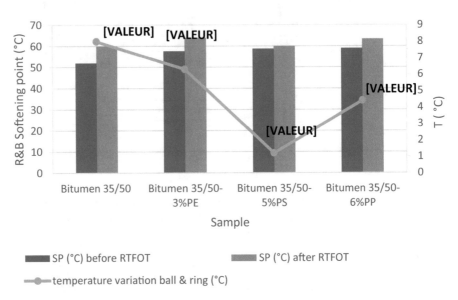

Fig. 9. Results of softening point before and after RTFOT

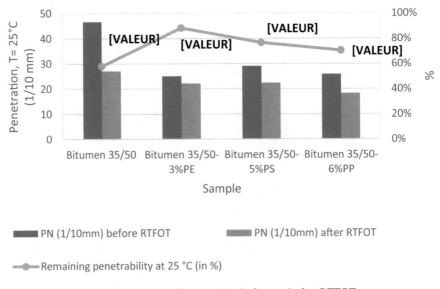

Fig. 10. Results of penetration before and after RTFOT

These results indicate that the addition of the recycled plastic waste has a significant impact on the characteristics of the base bitumen, tending to reduce the effect of age hardening (Ragni et al. 2018) and help prepare more durable asphalt pavement concrete (Nouali et al. 2019).

4 Conclusions

This study was conducted to investigate the effect of three types of recycled plastics, low-density polyethylene (LDPE), polypropylene (PP) and polystyrene on the conventional properties, temperature sensitivity and rheological properties of pure bitumen 35/50. The results obtained shows that the addition of recycled plastic waste in 35/50 grade bitumen reduces the penetration at 25 °C of RPMB by more than 50% and increases the softening point temperature by more than 15%. These variations indicate, on one hand, that our bitumen is increasingly resistant to excessive temperature variation and on the other hand, the interaction that happened between the polymers and the molecules of the pure bitumen. To summarize, by using recycled plastic waste we could increase the pure bitumen grade from 35/50 to 20/30 and create a binder that will enhance the pavement structure resistance to deformation at high temperatures and high resistance to deformation.

The penetration index values and bitumen aging results indicate that the addition of recycled plastics reduces the temperature susceptibility of RPMB except for RPMB at $\geq 5\%$ LDPE which becomes brittle with high elastic properties and more sensibility to low temperatures. Therefore, the mixtures that presented the best properties were RPMB at 3% LDPE, RPMB at 5% PS and RPMB at 6% PP. Also, according to the FTIR spectra we can say that the modification was mainly a physical process that preserves the pure bitumen structure and confirms the success of the RPMB mixture.

At term, this study showed that the incorporation of plastic waste into pure bitumen represents an alternative recycling method for plastic waste management that reduces the consumption rate of pure bitumen in Morocco and improves the binder's properties. In addition, further studies should be carried on to analyze and resolve the critical issue of RPMBs storage stability.

Acknowledgments. The authors would like to acknowledge the Technical Centre for Plastics and Rubber (CTPC) and Public laboratory for tests and studies - Center for Studies and Research of Transport Infrastructures (CERIT), Morocco, for providing opportunities for experimental collaboration, we also thank to BITUMA company for providing research facilities.

References

Abdullah, M.E., Abd Kader, S.A., Putra Jaya, R., et al.: Effect of waste plastic as bitumen modified in asphalt mixture. In: MATEC Web Conference, vol. 103, p. 09018 (2017a). https://doi.org/10.1051/matecconf/201710309018

Abdullah, M.E., Ahmad, N.A., Jaya, R.P., et al.: Effects of waste plastic on the physical and rheological properties of bitumen. In: IOP Conference Series: Materials Science and Engineering, vol. 204, p. 012016 (2017b). https://doi.org/10.1088/1757-899x/204/1/012016

Airey, G.D.: Factors affecting the rheology of polymer modified bitumen (PMB). In: Polymer Modified Bitumen, pp. 238–263. Elsevier (2011)

Alataş, T., Yilmaz, M.: Effects of different polymers on mechanical properties of bituminous binders and hot mixtures. Constr. Build. Mater. **42**, 161–167 (2013). https://doi.org/10.1016/j.conbuildmat.2013.01.027

Alataş, T., Yılmaz, M., Kök, B.V., Koral, A.f.: Comparison of permanent deformation and fatigue resistance of hot mix asphalts prepared with the same performance grade binders. Constr. Build. Mater. **30**, 66–72 (2012). https://doi.org/10.1016/j.conbuildmat.2011.12.021

Appiah, J.K., Berko-Boateng, V.N., Tagbor, T.A.: Use of waste plastic materials for road construction in Ghana. Case Stud. Constr. Mater. **6**, 1–7 (2017). https://doi.org/10.1016/j.cscm.2016.11.001

Asphalt Institute, European Bitumen Association: The bitumen industry: a global perspective: production, chemistry use, specification and occupational exposure (2015)

Blazejowski, K., Dolzycki, B.: Relationships between asphalt mix rutting resistance and MSCR test results. In: Design, Analysis, and Asphalt Material Characterization for Road and Airfield Pavements, Yichang, Hubei, China, pp. 202–209. American Society of Civil Engineers (2014)

Brasileiro, L.L., Moreno-Navarro, F., Martínez, R.T., et al.: Study of the feasability of producing modified asphalt bitumens using flakes made from recycled polymers. Constr. Build. Mater. **208**, 269–282 (2019). https://doi.org/10.1016/j.conbuildmat.2019.02.095

Carreau, P.J., Bousmina, M., Bonniot, F.: The viscoelastic properties of polymer-modified asphalts. Can. J. Chem. Eng. **78**, 495–503 (2000). https://doi.org/10.1002/cjce.5450780308

French Chamber of Commerce and Industry of Morocco (CFCIM) (2014) Waste-Morocco

Habib, N.Z., Kamaruddin, I., Napiah, M., Tan, I.M.: Rheological properties of polyethylene and polypropylene modified bitumen. Int. J. Civ. Environ. Eng. **4**, 5 (2010)

Hunter, R.N., Self, A., Read, J., Shell International Petroleum Co. (eds.): The Shell Bitumen Handbook, 6th edn. ICE Publishing/Telford, London (2015)

Isacsson, U., Zeng, H.: Relationships between bitumen chemistry and low temperature behaviour of asphalt. Constr. Build. Mater. **11**, 83–91 (1997). https://doi.org/10.1016/S0950-0618(97)00008-1

John, R., David, W.: The Shell Bitumen Handbook, 5th edn. (2003)

Köfteci, S., Ahmedzade, P., Kultayev, B.: Performance evaluation of bitumen modified by various types of waste plastics. Constr. Build. Mater. **73**, 592–602 (2014). https://doi.org/10.1016/j.conbuildmat.2014.09.067

Ministry of Equipment, Transport, Logistics and Water, Morocco: Road network of the kingdom (2019). www.equipement.gov.ma. http://www.equipement.gov.ma/routier/Infrastructures-Routieres/Reseau-Routier-du-Royaume/Pages/Importance-du-reseau.aspx. Accessed 30 May 2019

Nouali, M., Derriche, Z., Ghorbel, E., Chuanqiang, L.: Plastic bag waste modified bitumen a possible solution to the Algerian road pavements. Road Mater. Pavement Des. 1–13 (2019). https://doi.org/10.1080/14680629.2018.1560355

Plastics Europe: Plastics – the Facts : an analysis of European plastics production, demand and waste data (2017)

Polacco, G., Berlincioni, S., Biondi, D., et al.: Asphalt modification with different polyethylene-based polymers. Eur. Polym. J. **41**, 2831–2844 (2005). https://doi.org/10.1016/j.eurpolymj.2005.05.034

Polacco, G., Filippi, S., Merusi, F., Stastna, G.: A review of the fundamentals of polymer-modified asphalts: asphalt/polymer interactions and principles of compatibility. Adv. Colloid Interface Sci. **224**, 72–112 (2015). https://doi.org/10.1016/j.cis.2015.07.010

Ragni, D., Ferrotti, G., Lu, X., Canestrari, F.: Effect of temperature and chemical additives on the short-term ageing of polymer modified bitumen for WMA. Mater. Des. **160**, 514–526 (2018). https://doi.org/10.1016/j.matdes.2018.09.042

Singh, N., Hui, D., Singh, R., et al.: Recycling of plastic solid waste: a state of art review and future applications. Compos. Part B Eng. **115**, 409–422 (2017). https://doi.org/10.1016/j.compositesb.2016.09.013

Tauste, R., Moreno-Navarro, F., Sol-Sánchez, M., Rubio-Gámez, M.C.: Understanding the bitumen ageing phenomenon: a review. Constr. Build. Mater. **192**, 593–609 (2018). https://doi.org/10.1016/j.conbuildmat.2018.10.169

Tayfur, S., Ozen, H., Aksoy, A.: Investigation of rutting performance of asphalt mixtures containing polymer modifiers. Constr. Build. Mater. **21**, 328–337 (2007). https://doi.org/10.1016/j.conbuildmat.2005.08.014

Tsantilis, L., Dalmazzo, D., Baglieri, O., Santagata, E.: Effect of SBS molecular structure on the rheological properties of ternary nanomodified bituminous binders. Constr. Build. Mater. **222**, 183–192 (2019). https://doi.org/10.1016/j.conbuildmat.2019.06.095

Van den Bergh, W.: The effect of ageing on the fatigue and healing properties of bituminous mortars. [s.n.] (2011)

Wang, T., Yi, T., Yuzhen, Z.: The compatibility of SBS-modified asphalt. Pet. Sci. Technol. **28**, 764–772 (2010). https://doi.org/10.1080/10916460902937026

Weigel, S., Stephan, D.: The prediction of bitumen properties based on FTIR and multivariate analysis methods. Fuel **208**, 655–661 (2017). https://doi.org/10.1016/j.fuel.2017.07.048

Yao, Z., Zhang, J., Gao, F., et al.: Integrated utilization of recycled crumb rubber and polyethylene for enhancing the performance of modified bitumen. Constr. Build. Mater. **170**, 217–224 (2018). https://doi.org/10.1016/j.conbuildmat.2018.03.080

Zhu, J., Birgisson, B., Kringos, N.: Polymer modification of bitumen: advances and challenges. Eur. Polym. J. **54**, 18–38 (2014). https://doi.org/10.1016/j.eurpolymj.2014.02.005

Stress-Strain Behavior of Municipal Solid Waste Using Constitutive Modeling Approach

Sandeep K. Chouksey$^{(\boxtimes)}$

Department of Civil Engineering, National Institute of Technology, Raipur, India
schouksey.ce@nitrr.ac.in

Abstract. A generalized constitutive model is proposed to predict the stress-strain behavior of municipal solid waste in undrained loading condition, based on the critical state soil mechanics framework that incorporates the effects of time dependent mechanical creep and time-dependent biodegradation. The model parameters are calculated based on laboratory one-dimensional compression and triaxial compression tests. The validation of proposed model is presented in terms of stress-strain response and compared with experimental as well as with hyperbolic model. The predicted results from proposed model give good agreement with experimental results.

1 Introduction

The stress-strain response of MSW provides understanding of (i) role of compaction of MSW in settlement; (ii) to understand the deformation related issues leading to instability; (iii) role of leachate generation and drainage in the stability of waste fill. Prediction of stress-strain response and volume changes will give an idea of stress at failure state. In the literature, several authors reported triaxial compression tests under drained and undrained loading, contributions of Babu et al. (2010, a, b), Chouksey et al. (2013) and Chouksey and Babu (2013) are some of the studies in this direction. In most of the cases, the stress-strain response of MSW shows continuous increase of strain without showing any peak stress or ultimate failure stress. The objective of this paper is to describe the stress-strain response of MSW in drained condition using constitutive modeling approach. In the present study, a generalized constitutive model is developed based on critical state soil mechanics framework that accounts elastic, plastic, creep and biodegradation effects. The newly developed yield surface is derived with consideration of five parameters. The following sections present the development of the model, and validation with reference to the experimental data.

2 Proposed Constitutive Model

The mechanical behavior follows elasto-plastic behavior in the framework of critical state soil model with associated flow rule; The secondary compression is governed by the time dependent phenomenon in exponential function similar to the assumption of Gibson and Lo's (1961) model which is given by

© Springer Nature Switzerland AG 2020
H. Ameen et al. (Eds.): GeoMEast 2019, SUCI, pp. 146–149, 2020.
https://doi.org/10.1007/978-3-030-34199-2_10

$$\varepsilon_v^c = -b\sigma_{11}'\left(1 - e^{-ct'}\right) \tag{1}$$

where, b is the coefficient of mechanical creep; $\Delta\sigma 11$ is the change in mean effective stress, c is the rate constant for mechanical creep; and t' is the time since application of the stress increment and negative sign indicate that decrease in void ratio with time increases.

The biological composition is related to time and the total amount of strain that can occur due to biological decomposition. The time dependent biological degradation is proposed by Park and Lee (1997) and which is given by

$$\varepsilon_v^b = -E_{dg}(1 - e^{-dt''}) \tag{2}$$

Where, E_{dg} is the total amount of strain that can occur due to biological decomposition; d is the rate constant for biological decomposition; and t'' is the time since placement of the waste in the landfill.

2.1 Strain Due to Mechanical Creep and Biodegradation

In addition to elastic and plastic behavior of the municipal solid waste, considering compression due to mechanical creep and biological decomposition, the total volumetric strain of the MSW can be expressed as:

Total strain = volumetric strain due to elastic + volumetric strain due to plastic + volumetric strain due to creep and volumetric strain due to biodegradation.

$$\varepsilon_v = \varepsilon_v^e + \varepsilon_v^p + \varepsilon_v^c + \varepsilon_v^b \tag{3}$$

Substituting all the values of strains in Eq. (3). The final expression for the yield surface is obtained as:

$$q = Mp'\sqrt{\left(\frac{p_0'}{p'}\right)\exp\left[\left(\frac{e_0 - e}{1 + e_0} + b\sigma_{11}'(1 - e^{-ct}) + E_{dg}(1 - e^{-ct})\right)\frac{1 + e_0}{\lambda}\right] - 1} \tag{4}$$

where, p' is the mean effective stress and p_0' is the pre- consolidation pressure. The Eq. (4) is the proposed new model for MSW, which is an extended form of modified cam clay model that predicts the stress-strain behavior under different loading conditions. In addition to elastic and plastic strains, the total volumetric strain includes mechanical compression under loading as well as mechanical creep and biological degradation effects represented by Eqs. (1) and (2). M is the frictional constant, is the initial void ratio, and e is the void ratio after load increment and compressibility parameters, λ and κ are obtained from one dimensional consolidation tests.

3 Results and Discussions

The results are presented in terms of comparison of stress-strain responses using experimental, prediction from proposed model as well as from hyperbolic model. Stress-strain response shown in Fig. 1 corresponding to MSW sample of unit weight of 7.4 kN/m^3 show that deviatoric stress increases continuously without showing any sign of failure or ultimate stress. The deviatoric stress is calculated using Eq. (4) and compared with experimental results. The input parameters used for the validation of model is summarized in Tables 1 and 2. Similar results are obtained for other confining pressures.

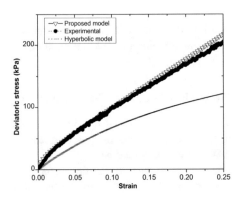

Fig. 1. Comparison of stress-strain response of MSW (for unit weight of 7.4 kN/m^3 at confining pressure of 50 kPa)

Table 1.

$\gamma(kN/m^3)$	λ	κ	ϕ'	p'_0 (kPa)	p'_c (kPa)	v
7.4	0.25	0.015	27.6	50,100 & 150	50,100 & 150	0.40

Table 2.

b (m^2/kN)	c (day^{-1})	d (day^{-1})	E_{dg}
0.00070	0.0069	0.00027	0.218

The experimental results are also compared with the prediction obtained from hyperbolic model. This nonlinear behavior observed during shearing of MSW in drained loading is captured by proposed model with combination of additional strains due to time dependent mechanical creep and biodegradation effects. It is noted that the predicted values of deviatoric stress calculated from hyperbolic model give good predictions up to 10% of strain, and asymptotic behavior is observed subsequently whereas predicted results obtained from proposed model capture the trend of stress-strain response of MSW adequately up to the higher percentages strain.

This agreement is attributed to the main feature of the proposed constitutive model that the critical state of soil concept has been extended to include mechanical creep and biological degradation effects of MSW. For example, deviatoric stress at 25% strain was observed 205 kPa whereas, at same strain, deviatoric stress evaluated using hyperbolic and proposed model was 121 and 213 kPa respectively. The hyperbolic model calculates 40% less deviatoric stress and proposed model gives 3.9% higher value compared to measured values of deviatoric stress.

4   Conclusions

In this paper, a generalized constitutive model for MSW is proposed based on critical state soil mechanics framework. The model accounts for strains due to elastic, plastic, creep and biodegradation effects. The following major conclusions can made:

1. The predicted values of deviatoric stress from proposed model are in good agreement with the experimental results for lower as well as at higher strains for all confining pressures whereas the predicted values from hyperbolic model gives good prediction up to 12% of strain.
2. The model parameters are evaluated from triaxial compession and one dimensional compression tests.

References

Babu, G.L.S., Reddy, K.R. Chouksey, S.K.: Constitutive model for MSW considering creep and biodegradation effects. In: 6ICEG International Conference at New Delhi, November, pp. 451–456 (2010)

Babu, G.L.S., Reddy, K.R., Chouksey, S.K.: Constitutive model for municipal solid waste incorporating mechanical creep and biodegradation-induced compression. Waste Manage. J. **30**(1), 11–22 (2010a)

Babu, G.L.S., Reddy, K.R., Chouksey, S.K.: Parametric study of MSW landfill settlement model. Waste Manag. J. **31**(1), 1222–1231 (2011)

Babu, G.L.S., Reddy, K.R., Chouksey, S.K., Kulkarni, H.: Prediction of long-term municipal solid waste landfill settlement using constitutive model. In: Practice Periodical of Hazardous, Toxic, and Radioactive Waste Management, ASCE, vol. 14, no. 2, pp. 139–150 (2010b)

Chouksey, S.K., Babu, G.S.: Constitutive model for strength characteristics of municipal solid waste. Int. J. Geomech. **15**(2), 04014040 (2013)

Babu, G.S., Chouksey, S.K., Reddy, K.R.: Approach for the use of MSW settlement predictions in the assessment of landfill capacity based on reliability analysis. Waste Manage. **33**(10), 2029–2034 (2013)

Gibson, R.E., Lo, K.Y.: A theory of soils exhibiting secondary compression. ActaPolytech. Scand. **10**, 1–15 (1961)

Park, H.I., Lee, S.R.: Long-term settlement behavior of landfills with refuse decomposition. J. Resour. Manage. Technol. **24**(4), 159–165 (1997)

Author Index